TROPICAL MARINE ENVIRONMENTS

Ruth Searle

Tropical Marine Environments

Evolution and ecology in the oceans

Written and illustrated by
Ruth Searle PhD

First published in Great Britain in 2014

Holly Blue Publishing
www.hollybluepublishing.co.uk

ISBN-13: 978-1503124325

ISBN-10: 1503124320

Contents

Tropical marine environments

Coral reefs are one of the most beautiful and diverse marine habitats on Earth, they cover around 300,000 square km of the world's oceans and support an amazing collection of animals and plants. Although the tropical seas are generally unproductive areas, the reef ecosystem is completely self-contained and recycles nutrients in a highly efficient way so that primary productivity is between 30 and 250 times that of the open ocean. Around 80% of the plankton suspended in the water is taken out by corals and other suspension feeders and in sandy areas, deposit feeders such as sea cucumbers ingest the detritus that accumulates in the sediment. Grazers such as molluscs and sea urchins feed on the algae growing on hard substrates and fish such as the parrotfish, *Scarus* sp. feed on corals, crushing the hard calcareous skeletons using its plate-like teeth. After digesting the polyps, the skeleton is excreted as fine sand. Carnivores eat smaller fish and the whole reef system forms a complex food web.

The structure of corals

Corals belong to the phylum Cnidaria, which, you may remember, include jellyfish and sea anemones. Within this phylum is the class Anthozoa, to which both anemones and corals belong. There are more than 6,000 species of anthozoans and it is the largest class of cnidarians. The anthozoans are all polypoid - that is they have the structure of a polyp, with a gastrovascular cavity and tentacles.

Corals belong to the class Anthozoa within the phylum Cnidaria. There are over 6,000 species.

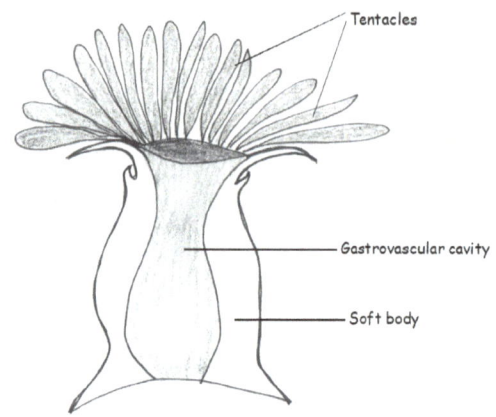

Tentacles

Gastrovascular cavity

Soft body

Anemone

Fig 1. Class Anthozoa. An anemone

Corals are divided into two main groups, the scleractinian or stony corals and the octocorals or soft corals.

The difference between anemones, seen in figure 1, and coral polyps, figure 2, is that anemones do not have an external skeleton as corals do. Corals have an exoskeleton of calcium carbonate, or limestone that builds the spectacular coral reefs of the tropical marine environment. Reef building corals are called hermatypic corals. However, not all coral animals build reefs and some solitary and colonial animals live in deeper, darker water and in colder conditions than most reef building corals.

Fig 2. Cnidaria. A coral polyp showing a cross section through the coral cup.

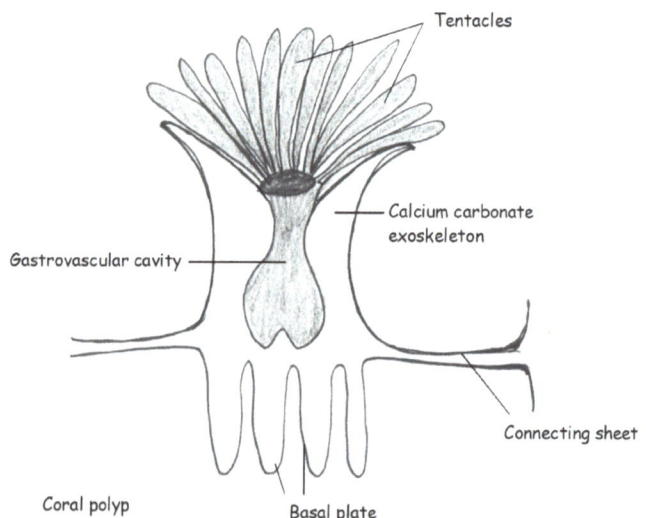

Tentacles

Calcium carbonate exoskeleton

Gastrovascular cavity

Connecting sheet

Coral polyp

Basal plate

~~~~RESEARCH HIGHLIGHTS~~~~

Coral cavities are sinks of dissolved organic carbon
The cavities within coral polyps have been found to act as sinks for dissolved organic carbon (DOC) and along with their associated bacterioplankton, play an important role in the energy budget of the reef. At depths of 5-17 m, coral reefs off the coast of Venezuela and Indonesia were studied and researchers found that the cavities within the coral polyps reduce DOC by two orders of magnitude over and above that of their associated bacteria. This had important implications for global warming and the uptake of dissolved carbon – we can't afford to destroy our coral reefs!
(de Goeij et al. 2007)

Sea anemones

Sea anemones are the largest of the anthozoans ranging from one to several centimeters in diameter, although one large species on the Great Barrier Reef is around 1 m in diameter. Most live on the hard substrates of the reef and each anemone has stinging tentacles called nematocysts, which are used to capture prey and to defend themselves. Some fish and shrimps live among the nematocysts, protected from the stinging cells by the

Fig 3. The clownfish. *Amphiprion*, among the stinging tentacles of a sea anemone.

~~~~RESEARCH HIGHLIGHTS~~~~

Nutrient transfer between anemone fish and their host anemones
Anemone fish, such as *Amphiprion bicinctus* excrete ammonia, especially just after feeding. Researchers have now shown through laboratory experiments that the host anemone takes up this ammonia via photosynthesizing zooxanthellae. During the night, when photosynthesis stops, the absorption of ammonia dropped to zero but during the day, when zooxanthellae are most active, practically all the ammonia produced by the anemone fish is absorbed rapidly by the anemone. This symbiosis is an important aspect of coral reef productivity.
(Roopin et al. 2008)

production of mucous. Figure 3 shows an anemone fish, the clownfish, *Amphiprion*, among the tentacles of an anemone, where it is protected from predators. The anemone benefits from its host by gaining scraps of food.

Anemones and coral polyps have stinging tentacles called nematocysts.

Corals

The corals are the individual animals that make up the reef and are divided into two main groups, the scleractinian or stony corals and the octocorals or soft corals. Scleractinian corals are the corals that are most associated with coral reefs, as we saw earlier. The polyps are smaller than most anemones and they are mostly found in colonies.

They secrete a calcium carbonate skeleton of about 1-3 mm in diameter from the base of the polyp called the basal plate. The corals are connected together by a sheet of calcium carbonate. Like anemones, coral polyps have stinging nematocysts.

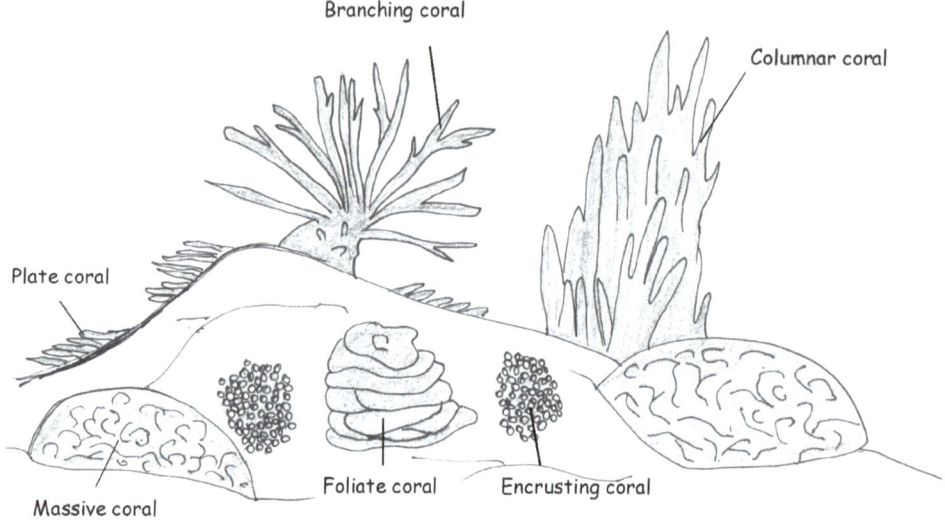

Fig 4. Corals. The various forms of scleractinian corals

Figure 4 illustrates how some scleractinian corals are low, encrust over the substrate and grow slowly at about 0.5 to 2 cm per year whilst others, such as the staghorn coral are higher, branching, and grow more quickly at about 10 cm per year. Scleractinian corals can reproduce asexually in a similar way to the anemones, via budding of new individuals and also sexually by producing planktonic larvae that disperse and produce new colonies.

There are differences in reproduction in the Atlantic and Indo West Pacific corals. For example, one study shows that reproduction in the Caribbean, at least in the dominant genus *Acropora*, is primarily asexual via fragmentation and budding. In the Pacific, recruitment is thought to be primarily sexual. Some species have long breeding seasons with few eggs and large larvae that are brooded by the parent. Others are oviparous with short breeding seasons eg; *Acropora*.

Hundreds of coral species reproduce in mass spawning events, where individuals are frequently hermaphrodites (both male and female). This spectacular event appears to be related to the lunar cycle (see Research Highlights right). Hybridization among mass spawning species in the Caribbean is common and some rare Indo-Pacific coral species of Acropora are probably hybrids and this may render them less

~~~~RESEARCH HIGHLIGHTS~~~~

Mass spawning
At three different locations, 32 species of coral in the Great Barrier Reef undertake mass spawning. They are usually hermaphrodites (individuals are both male and female), spawning together 5-8 nights after a full moon in October or November, although there is some variability depending on conditions. Recruitment rates (numbers of new corals) have been found to vary substantially among locations on the Great Barrier Reef and do not match the variation in the abundance of adults (Harrison et al. 1984). Spawning does not correspond to tidal movements and is thought to be synchronized by lunar cycles (Babcock et al.1994). Through genetic studies, mass spawning corals have also been shown to hybridize frequently
(Hatta et al.1999; Richards et al.2008).

10

vulnerable to extinction. Mass spawning also provides valuable nutrition in the form of spawn-derived nitrogen (N) and carbon (C) which is recycled within the reef system and can trigger an intense bloom of benthic dinoflagellates.

The octocorals, shown in figure 5, include the gorgonian and the soft corals. Gorgonian corals are commonly found on the Indo West Pacific reefs and include the sea pens, sea whips and sea fans. These corals are designed to bend in the currents and their branching rods collect planktonic food. Some grow to over 2 m in height.

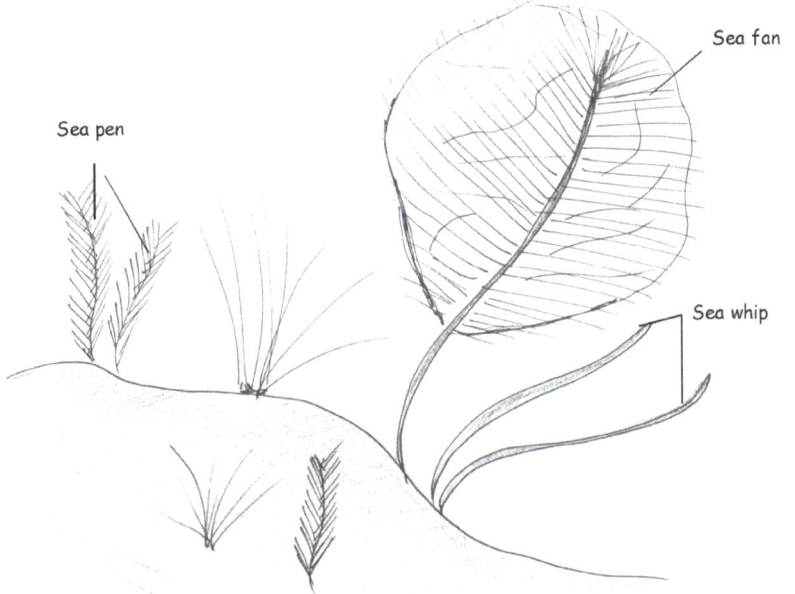

Fig 5. Soft corals. The gorgonian or soft corals which are more common at depth than scleractinian corals.

Zooxanthellae

Most reef building corals have a symbiotic relationship with dinoflagellates known as zooxanthellae, shown in figure 6. If you remember, we met dinoflagellates as members of the phytoplankton. Here they live in a symbiotic relationship within coral polyps and have lost their flagellum (tail). Many corals have a yellowish brown colour because of the presence of certain zooxanthellae. The most common species of zooxanthellae is the dinoflagellate, *Symbiodinum microadriaticum,* although, so far eight lineages of zooxanthellae have been identified using nuclear ribosomal DNA and chloroplast DNA, with each lineage containing many species. Most coralcolonies appear to associate with a single zooxanthellae type but there are some corals that can associate with several types simultaneously.

> Reef building corals have a symbiotic relationship with zooxanthellae. Coral bleaching occurs when zooxanthellae are expelled from the coral tissue.

The majority of corals in the Great Barrier Reef associate with just one type while there is a second type present at low levels. Different combinations of symbionts may provide the coral with advantages within different ecological niches and may affect their ability to adapt to changing conditions, such as increases in sea temperature and light intensity.

~~~~RESEARCH HIGHLIGHTS~~~~

The significance of coral colours

The coral *Montipora monasteriata* occurs in several colours – tan, blue, brown, green and red within its range of depth related habitats – but little is known about the ecological and physiological significance of the different colours. Recently, researchers have found that pigments in the coral host respond to changes in irradiance, or light intensity, and that this helped symbiotic bacteria to retain chlorophyll. In turn, this enhances photosynthetic activity. This effect was seen most strongly in orange absorbing pigment. The implication is that when blue absorbing pigments bleach, or lose their symbiotic bacteria because of increased light intensity, the orange pigments will continue to survive.
(Dove et al. 2008)

Zooxanthellae live in the vacuoles within the cells of the coral polyp in very high densities. There may be more than 50 dinoflagellates per cell (more than 10^6 cm^{-2}) and each zooxanthellae produces over 90% of the nutrition required by the coral. This is achieved via photosynthesis, so corals need to remain in clear shallow water within the photic zone.

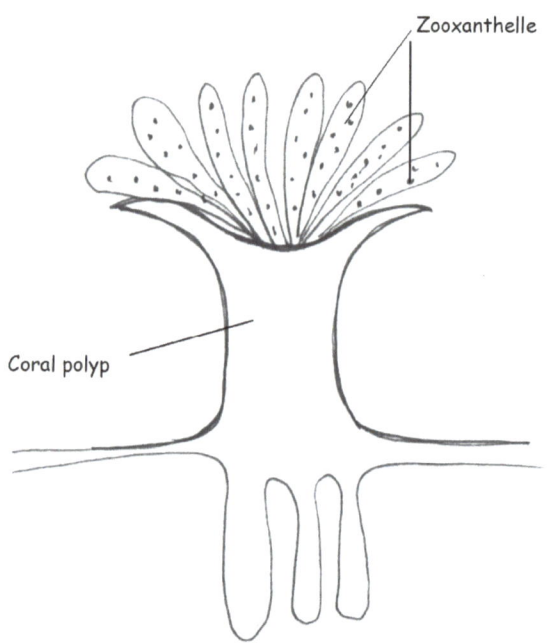

This symbiotic relationship is highly efficient at recycling nutrients within the coral reef system. Figure 7, below, shows how the coral receives food produced from photosynthesis and how the nitrogenous waste products of metabolism from the coral are used by the zooxanthellae. Zooxanthellae also gain protection within the coral polyp.

Fig 6. Zooxanthellae. Coral polyp showing symbiotic zooxanthellae.

Coral reefs are areas of high primary productivity because of this symbiotic relationship and are subsequently rich in animal diversity, with an estimated 500,000 different species. Coral bleaching is caused when the coral host expels the zooxanthellae, often caused by high sea temperatures.

Through photosynthesis, Zooxanthellae produce over 90% of the nutrition required by the coral.

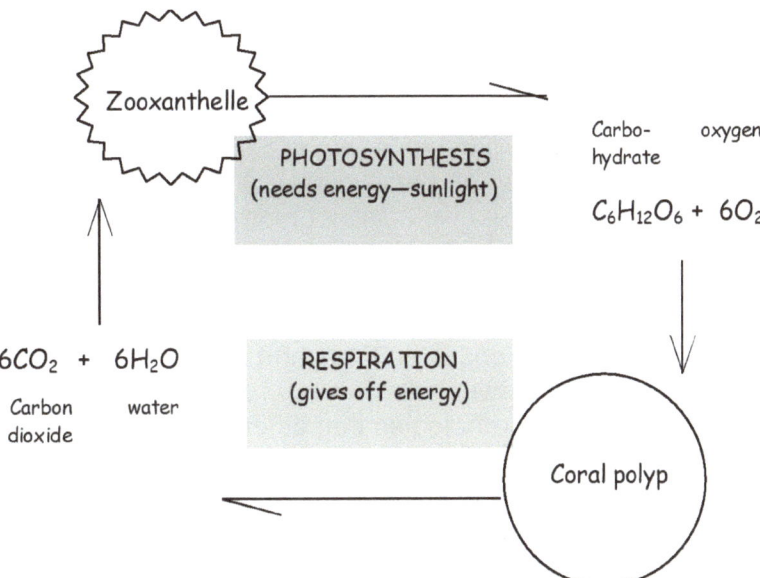

Fig 7. The symbiosis of corals and zooxanthellae. Zooxanthellae exchange the products of photosynthesis for the products of metabolism as this schematic diagram shows.

~~~~RESEARCH HIGHLIGHTS~~~~

Zooxanthellae and thermal tolerance in coral reefs
It is debatable whether coral reefs will survive the projected increases in temperature due to global warming. Will they adapt and acclimatize over the next few decades, or will corals die out? The number of bleaching episodes due to high temperature stress has increased in recent years, sparking a number of research studies. One study investigated the potential of a common and widespread Indo-Pacific hard coral, *Acropora millepora*, to acclimatize to increased temperatures. Adult corals in certain circumstances are capable of acquiring increased thermal tolerance because of a change in the type of symbiotic bacteria inhabiting their tissues. *Symbiodinium* species of zooxanthellae occur together as varying types (C to D) within the tissues of the coral polyp and under thermal stress were shown to shuffle about so that a different type was dominant. No new bacteria were introduced from the surrounding water but type D became the dominant strain and are the most thermally tolerant. This gave the coral an increased tolerance to heat of 1-1.5°C. While this is good news and gives hope for the survival of coral reefs in the short-term, it may not be enough to ensure the survival of corals over the next 100 years, if predictions of increases in tropical sea surface temperature of 1-3°C are correct.
(Berkelmans and van Oppen 2006)

Hybridization and diversity on the reef

Coexisting on coral reefs are around 105 coral species from 36 genera and 11 families that reproduce annually in synchronous mass-spawning events and it is thought that hybridisation between species is common. In fact, laboratory experiments have shown that crossbreeding from a number of mass spawning genera produce viable hybrids. This interspecific hybridisation should reduce diversification, yet many corals that reproduce through mass spawning have rapidly diversified, confusing our understanding of coral evolution. They have fossil records dating back at least 3 to 3.6 My.

One of the most speciose coral groups in the world is *Acroper,* with 115 species having evolved over the past 5 million years, many capable of interspecific hybridisation in laboratory crosses. *Acropora* corals, *A. Cervicornis* and *A. palmata* are genetically distinct sister species with distinct morphologies and habitat preferences. The staghorn coral, *A. Cervicornis,* occurs in the forereef and back reef, while the Elkhorn coral, *A. Palmate,* occurs in the reef crest habitats where there is high wave energy. *Acropora prolifera* is a third species of *Acropora* that occurs throughout the Caribbean and is an F1 hybrid of these two species. Varieties of *A. prolifera* are intermediate between *A. Cervicornis* and *A. Palmata* and, as figure 8 (a to d) shows, have unique morphologies that differ depending on which species provides the egg for hybridisation. Although the potential for evolution of these hybrids is limited, individual *A. prolifera* or able to reproduce asexually and are long-lived.

The different morphologies of *A. prolifera* vary between a thin, highly branched form, known as the bushy morph (fig 8 c) and a thicker, more flattened form known as the palmate morph (fig 8 d).

Figure 8. The Caribbean *Acropora* species: (**A**) *A. cervicornis* and (**B**) *A. palmata*, and (**C**) the bushy and (**D**) palmate F1 hybrid *A. prolifera* morphs from Puerto Rico. (Taken from Steven V. Vollmer and Stephen R. Palumbi (2002) Hybridization and the Evolution of Reef Coral Diversity. Science. June 14th. Vol 296. No 5575, pp.2023 – 2025).

Acropora demonstrates that reef building corals not only diversify through conventional species formation but also through long-lived coral hybrids which breed asexually and generate new morphologies and potential new coral types without speciation. It remains to be seen how pervasive these hybrid corals are but it has been suggested that they may be common due to the potential for natural hybridisation in mass spawning corals.

The distribution and zonation of coral reefs

There are six major physical factors that limit coral reef development: temperature, depth, light, salinity, sedimentation and emersion into air.

~~~~RESEARCH HIGHLIGHTS~~~~

Experiments with coral transplants
Closely related scleractinian coral species, *Porites cylindrical* and *Porites rus*, which occur in shallow reef flats were transplanted onto three different substrates: metal grids over sand, live coral and dead coral. With the onset of high sea temperatures during the 1998 El Nino event, all transplanted corals showed immediate signs of bleaching and tissue death. The *P. rus* species coped better but overall, the survival rates for transplanted species on metal grids (35%) were higher than those on dead (6%) or living coral (22%).
(Yap 2004)

Temperature

Hermatypic corals are found in waters bounded by the 20^0C surface isotherm and the lower temperature for reef formation is 18^0C, so tropical corals require a narrow range of temperatures in order to thrive. Global warming is a real threat to these coral animals as they are already living at the upper limits of temperature that they are able to survive. Optimum tropical reef development occurs in mean temperatures of 23-25^{0}C, while some corals can tolerate temperatures as high as 40^0C, although some types of corals do thrive outside the tropics, for example in the Bahamas, Florida Keys and Bermuda. Reefs are absent from the West coasts of South and Central America and the West coast of Africa due to the cold upwelling induced by the Peru and Benguela Currents respectively.

Depth and light

Reefs are also limited by depth and do not form in water deeper than 50-100 m. Most grow in water of 25 m or less and are restricted to continental margins or islands. This is due to light limitation in zooxanthellae production and the light compensation appears to be the depth where light intensity is 1-2% of light at the surface. Zooxanthellae are the limiting

There are six major physical factors that limit coral reef development: temperature, depth, light, salinity, sedimentation and emersion into air.

factor in the distribution of corals to warm shallow water. As you will see in figure 8, at about 50 m depth, the slope drops down into the deep water of the ocean. In the Pacific, corals do not grow below about 50-60 m but in the Caribbean, probably due to greater light penetration, corals grow to about 100 m depth.

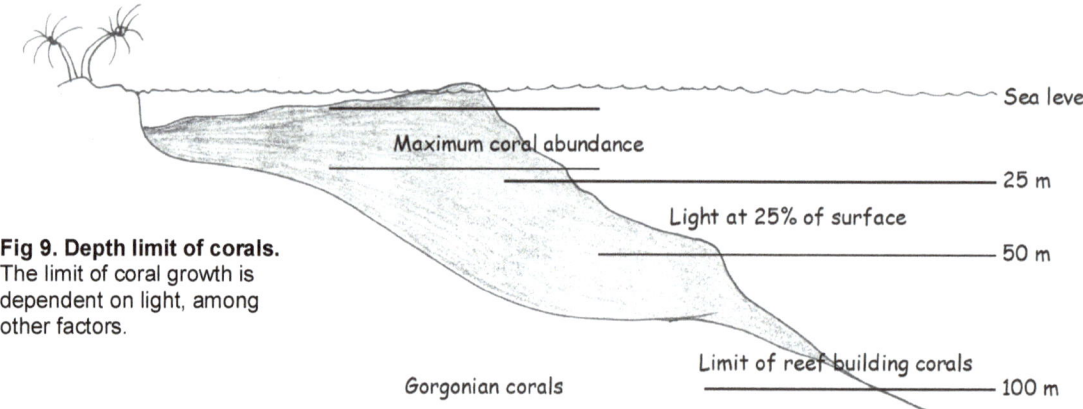

Fig 9. Depth limit of corals. The limit of coral growth is dependent on light, among other factors.

Salinity and sedimentation

Another limiting factor to coral growth is salinity. Where there is freshwater runoff, corals are missing, for example between the Amazon and Orinoco Rivers on the East coast of South America. This occurs on a smaller scale wherever there is freshwater runoff due to the reduction in salinity and also the associated increase in sedimentation that clogs feeding structures and reduces available light by turbidity or mixing in the water. Corals thrive in areas where there is wave action, which oxygenates the seawater, reduces sedimentation and delivers food in the form of plankton.

Emergence

Corals may survive emergence into air for a few hours at low tide, but generally their growth is limited to the level of the mean low water tides. Studies have shown 80-90% mortality in corals exposed by low tides over a 5 day period, so this is another significantly limiting factor.

~~~~RESEARCH HIGHLIGHTS~~~~

Sand transportation on a fringing reef
Recently, researchers studying a coral reef and adjacent beach at Molokai, Hawaii showed that sand grains are freely transported between the beach and the reef-flat, but that there is limited exchange of sand between the fore reef and either the beach or the reef flat. They also showed that the calcium carbonate content of the bottom sediment (sand, silt and clay) increases with the distance from shore from 400 to 650 m. They remain relatively stable across the fore reef. The study shows that the wide reef flat and reef crest blocks the transport of sand to the fore reef. Any sediment introduced to the inner reef flat is not quickly dispersed seaward and settles for long periods of time. This can cause problems for corals by blocking feeding structures.
(Calhoun et al. 2008)

Distribution of corals worldwide

Coral reefs are found in three principal areas; the Caribbean, including the Bahamas and Florida Keys, The Red Sea, including Indian Ocean Islands such as the

Seychelles and the Indo West Pacific. The centre of diversity is in the Indo West Pacific with attenuation in all directions.

Corals are generally found in a broad band throughout the tropics with extensions where there are warm water currents, eg; both on the West and East coasts of

> Coral reefs have a geological history stretching back over 500 million years and are amongst the most diverse and complex ecosystems on Earth.

Australia due to warm currents from the North, and as far as the southernmost islands of Japan. Belize also has a large area of reef. They are the result of growth since the last ice age, 10-11,000 yrs ago, although they have much older foundations, particularly the atolls. The following schematic map, figure 10, shows the distribution of corals worldwide between the 20^0C isotherms.

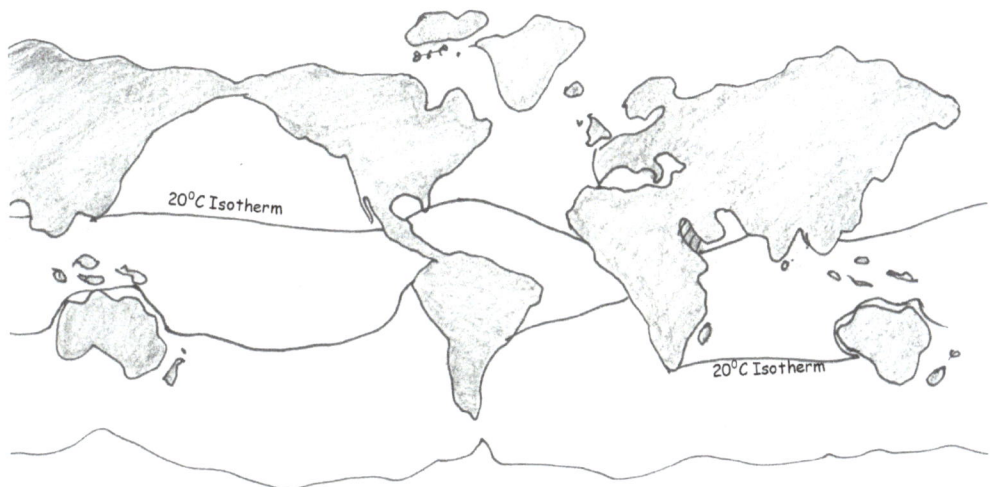

Fig 10. The distribution of coral reefs worldwide. Corals require a narrow temperature range in order to thrive and mainly grow between the 20^0C isotherms.

Reef formation

There are 4 types of reef: fringing reefs, barrier reefs, patch reefs and atolls. All are part of a series of forms which all develop in the same way. Where conditions are suitable, corals will grow in shallow clear water to about 45 m depths along tropical rocky coasts where there is no freshwater runoff. They

Fig 11. A South Pacific atoll. Lagoon and reef, as seen from the air.

grow upward until limited by emersion into air and begin to spread outward from the coast. Atolls are generally found in the Indo Pacific area, whilst fringing reefs and barrier reefs are found along continental coastlines and Islands such as Hawaii and throughout the tropical reef zones. They tend to grade into one another.

Atolls form when a submarine volcano develops a fringing reef, and as it sinks over time, the coral grow upward. The top of the guyot then subsides and a deep lagoon is formed in the centre of a ring of coral reefs.

Barrier reefs are found further offshore with a wider lagoon than fringing reefs and may indicate past increases in sea level. Patch reefs are fairly oval in shape in the axis of the prevailing winds, they may have a sandy cay on the leeward side and their size varies enormously. In the shelter of the Great Barrier Reef they become low wooded islands. Atolls tend to be oceanic and fairly circular with a series of sand cays or motus enclosing a deep lagoon. Their seaward slopes descend into deep water and there is a definite zonation through the atoll.

There are various theories to explain the origin of atolls, the first by Darwin in 1842, which is still accepted today. Borings at Enewetak and Bikini atolls reached basalt and limestone respectively and proved Darwin's theory. Further support came from the discovery of flat topped guyots, which are atolls which have been flattened by wave action.

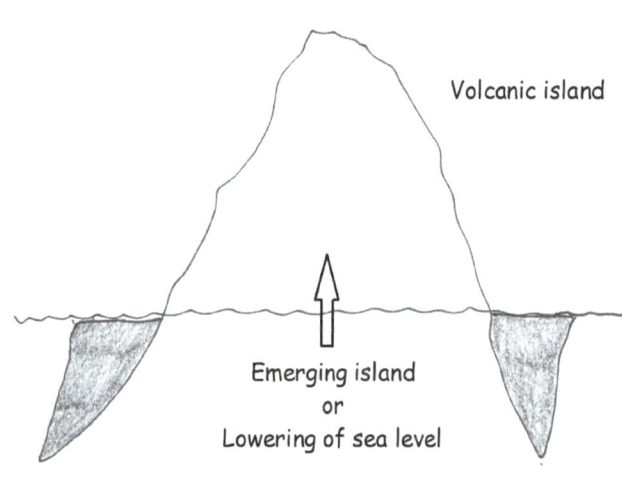

Atolls form when a submarine volcano, usually over hot spots in the tropical Pacific, develops a fringing reef, and as it sinks over time, the coral grow upwards to stay in the sunlight for zooxanthellae production. The top of the guyot subsides and a deep lagoon is formed in the centre of a ring of coral reefs, often with sandy cays and vegetation.

Fig 12. Atoll formation. Stage one, coral grows at the sides of an emerging volcano or when sea levels are lowered around an existing land mass.

Fig 13. Atoll formation
Stage two, corals continue to grow around the land as is sinks or sea levels rise, growing upward to remain in the light for zooxanthellae to photosynthesise.

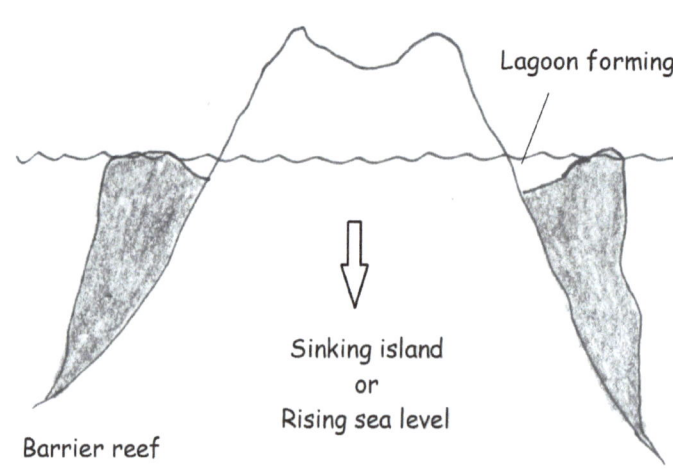

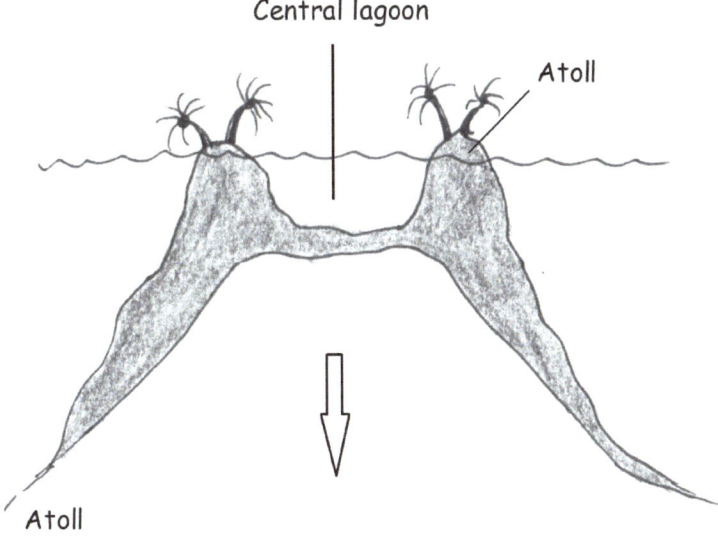

Central lagoon

Atoll

Atoll

Fig 14. Atoll formation
Stage three, as the land mass sinks further or sea levels continue to rise, a central lagoon forms surrounded by atolls.

There are 4 types of reef: fringing reefs, barrier reefs, patch reefs and atolls. All are part of a series of forms which all develop in the same way.

Zonation on the reef

Reefs display patterns or zones depending on the depth of the reef, wave action and exposure. Atolls have the most complex zonation of all. Locally, reefs may be divided up into more or fewer zones but a general schematic diagram, figure 15, shows the zonation of a reef.

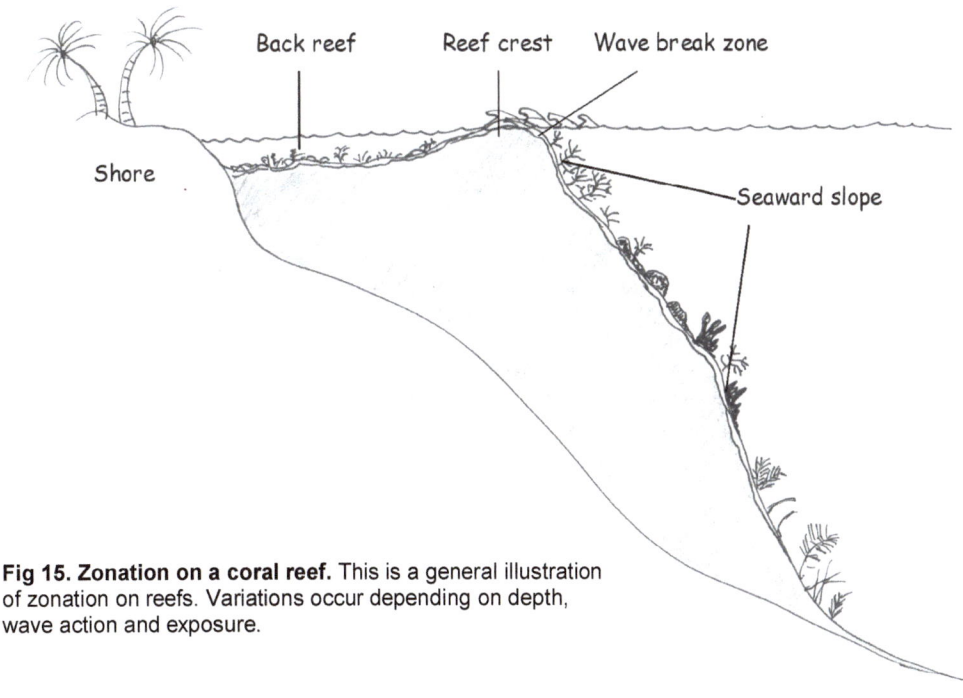

Back reef   Reef crest   Wave break zone

Shore

Seaward slope

Fig 15. Zonation on a coral reef. This is a general illustration of zonation on reefs. Variations occur depending on depth, wave action and exposure.

The back reef

The back reef is close to the shore on the sheltered side with a few rocky corals and possibly a fringe of mangroves. Then there is a narrow lagoon around 1-5 m deep to the seaward side. Large parts of it may be exposed at low tide. The back reef has a dominant biota of calcified green algae such as *Halimeda* along with various seagrasses such as *Thalassia* (turtle grass) and hummocks of coral such as *Porites*. It extends outward from the shore to the reef crest and may be anything from a few tens to a few thousand metres wide with a substrate of coral rock and loose sand.

> Reefs display patterns or zones depending on the depth of the reef, wave action and exposure.

The back reef is shallow and sheltered from wave action and as such, water circulation is limited and sediment tends to accumulate which contributes to poor conditions for coral growth. There are, however, many bottom dwelling species such as polychete worms, crabs and molluscs.

The reef crest

The reef crest lies on the outer side of the reef with its seaward side marked by a line of breaking waves. The reef crest is exposed at low tide and varies in width from a few metres to a few tens of metres. The reef crest is dominated by hardy corals that can withstand severe wave action but is restricted at the rear by sediment encroachment from the back reef. Species present on the reef crest include a few species of encrusting coral including *Porolithon, Hydrolithon* and *Lithothamnion*.

The wave break zone

The wave break zone is just to the seaward side of the reef crest and takes the full brunt of the waves. There is often a pattern of groves and deep channels here called the spur and groove zone, which forms because of exposure to constant wave action. The wave break zone is dominated with coral species such as *Acropora palmata* mixed with *Millepora*. In exposed areas such as Discovery Bay, Jamaica, which are subject to constant trade winds, there is a well defined spur and groove zone dominated by corals such as *Monastrea* and *Agricea*.

The seaward slope

The seaward slope extends from the low tide mark to deep water and it is here that many of the large corals thrive. The slope is dominated on the upper 15-25 m by large corals such as *Acropora cervicornis* and *Monastrea*. Below this, at 20-30 m depth, wave action is reduced and light is only about 25% of the light at the surface, so corals here tend to be the smaller branching types. At about 30-40 m depth, corals become patchy due to dwindling light and the accumulation of sediment. Gorgonian soft corals such as sea fans dominate.

Fig 16. The seaward slope. Gorgonean corals dominate at depths below 30-40 m.

The history of coral reefs

The coral reefs we see today date back just 5 million years ago to the Pliocene but the earliest reefs are thought to date to ~3.4 billion years ago to the early Archaean and were found in Strelley Pool Chert (SPC) (Pilbara Craton, Australia), a sedimentary rock formation containing stromatolites (see chapter X for a discussion about stromatolites). These early reefs were built by the metabolic activity of cyanobacteria along with other bacteria and because there were no predators on these microbial reefs, they grew large and complex over geological time. Figure 17 shows these stromatolite reef of the Strelley Pool Chert, Australia.

Fig 17. a–c, 'Encrusting/domical laminites'; d–f, 'small crested/conical laminites'; g–i, 'cuspate swales'; j–l, 'large complex cones' (dashed lines in k trace lamina shape and show outlines of intraclast conglomerate piled against the cone at two levels). m–o, 'Egg-carton laminites'; p, q, 'wavy laminites'; r–t, 'iron-rich laminites' (t is a cut slab). The scale card in b, h and i is 18 cm. The scale card increments in c, e, k, l, n and s are 1 cm. The scale bar in o is about 1 cm. The scale bars in the remaining pictures are about 5 cm (Taken from Allwood AC, Walter MR, Kamber BS, Marshall CP, Burch IW. 2006. Stromatolite reef from the Early Archaean era of Australia. *Nature* 441:714–18).

By the early Cambrian (~520 Mya) the earliest reefs built by invertebrate metazoans were found in the Pestrotsvet Formation (Tommotian) of the southeastern Siberian Platform. The most dominant reef building organisms were still microbes but metazoans, such as the *archaeocyaths*, a group of hypercalcifying sponges, built reefs in lime mudstone. Hypercalcification is the development of especially robust calcareous skeletons with a high skeletal-biomass ratio. Biohermss, as they are called, are intergrowths of *archaeocyaths, calcimicrobes (Renalcis)* and rare *coralomorphs (Cysticyathus)*.

> Hypercalcification is the development of especially robust calcareous skeletons with a high skeletal-biomass ratio.

They are meter-scale mounds found alone or stacked together and were found throughout the early Cambrian reef building phase in shallow water marine habitats. There is evidence that there was intense competition for space in these metazoan reefs and when the *archaeocyaths* became extinct, reef growth by microbes and a few siliceous sponges continued to dominate.

During the Ordovician, reefs building organisms diversified and reefs became more extensive.

During the Silurian and Devonian, there was very little change in reefs in terms of taxonomic composition, internal structure and biodiversity and reef building microbes and sponges, related to modern *Astrosclera* corals dominated the reefs.

At the end of the Devonian (375 Mya) there was a major collapse in metazoan reef building and during the subsequent periods of Carboniferous and Permian, reef building and the composition of reefs changed with the extinction of many tabulate corals, Palaeozoic stromatoporoids. In the early Carboniferous, reef building was dominated by microbes, rugose corals (a group of hypercalcifying bivalves with coral-like growth forms that proliferated during the Cretaceous period), bryozoans and some sponges. Later in the Carboniferous and early Permian, the dominant reef builders were calcareous algae which trapped mode between branches, rather than by forming a rigid, wave resistant framework.

Late in the Permian, a variety of hypercalcifying sponges, often associated with microbes, gained more prominence as reef builders. By the late Triassic, scleractinian corals became the most dominant and prolific reef builders. In the late Jurassic (~155 Mya) reefs were rich in hypercalcified sponges and microbial carbonates

> By the late Cretaceous (80 Mya) modern coral-coralline algal reefs had emerged and by the Pliocene, 5 Mya, modern reef distribution was established.

Coral reefs during the Cretaceous grew more slowly. Although rudest reefs were present, their reef building capacity was quite low and coral reefs had declined. There was also a mass extinction event at the end of the Cretaceous period, 65 Mya, further exacerbating the reduction in coral reefs. However, by the late Cretaceous (80 Mya) modern coral-coralline algal reefs had emerged.

The start of the Cenozoic era began with a significant rise in coral algal reefs, although they also declined by the end of the Paleocene, 56 Mya, and persisted throughout the Eocene.

Modern coral-algal reefs expanded throughout the Oligocene and me are seen and by the Pliocene, 5 Mya, modern reef distribution was established.

The diversity of coral reefs

Coral reefs have a geological history stretching back over 500 million years. The phyla have survived extremes of climate change throughout this time, from major ice ages in the Carboniferous and Permian to the more recent Quaternary. The greenhouse conditions of the Cretaceous were probably triggered by plate tectonics and changes in ocean circulation. Modern corals are amongst the most diverse and complex ecosystems on Earth, covering approximately 600,000 km^2, around 0.2% of the global ocean and about 15% of the shallow sea to a depth of about 30 m. The Great Barrier Reef is the largest, covering an area around 2,000 km by 145 km. This represents an incredible deposition of biologically produced calcium carbonate. As you may remember, the shells of plankton such as coccolithophores and foraminiferans form sediments of calcium carbonate on the sea floor which, when compressed, become limestone rock. Reef building corals contribute to the deposition of rock forming sediments.

> Coral reefs cover approximately 600,000 km^2, around 0.2% of the global ocean and about 15% of the shallow sea to a depth of about 30 m.

Marine biodiversity hotspots

Coral reefs around the world are being degraded worldwide by human activities (58%) and climate warming (25%). A study in 2002 analysed the geographic ranges of 3235 species from four phyla: fish, corals, snails and lobsters and showed that between 7.2% and 53.6% of each taxon have highly restricted ranges and are particularly vulnerable to extinction. The study also showed that restricted range species on tropical reefs are clustered into centres of endemism, like those described for terrestrial taxa. It found that 10 of the richest centres of endemism cover 15.8% of the world's coral reefs (0.012% of the oceans) but include between 44.8 and 54.2% of the restricted-range species. Worryingly many occur in regions where reefs are being severely degraded by humans. These centres of endemism are major biodiversity hotspots which desperately require effective conservation and management.

> The Great Barrier Reef is the largest coral reef in the world, covering an area around 2,000 km by 145 km. It can even be seen from space and is the largest structure ever made by living organisms.

Researchers working on this study mapped the geographic ranges of 1700 species of coral reef fish, 804 species of coral, 662 species of snail, and 69 species of lobster. Figure 18 (a to d) shows global clines in species richness for these taxa and the high level of concordance in patterns of total species richness, with a peak in the 'Coral Triangle' of south-east Asia, falling rapidly moving east across the Pacific and less rapidly moving west across the Indian Ocean. Species richness in the Atlantic is highest in the Caribbean.

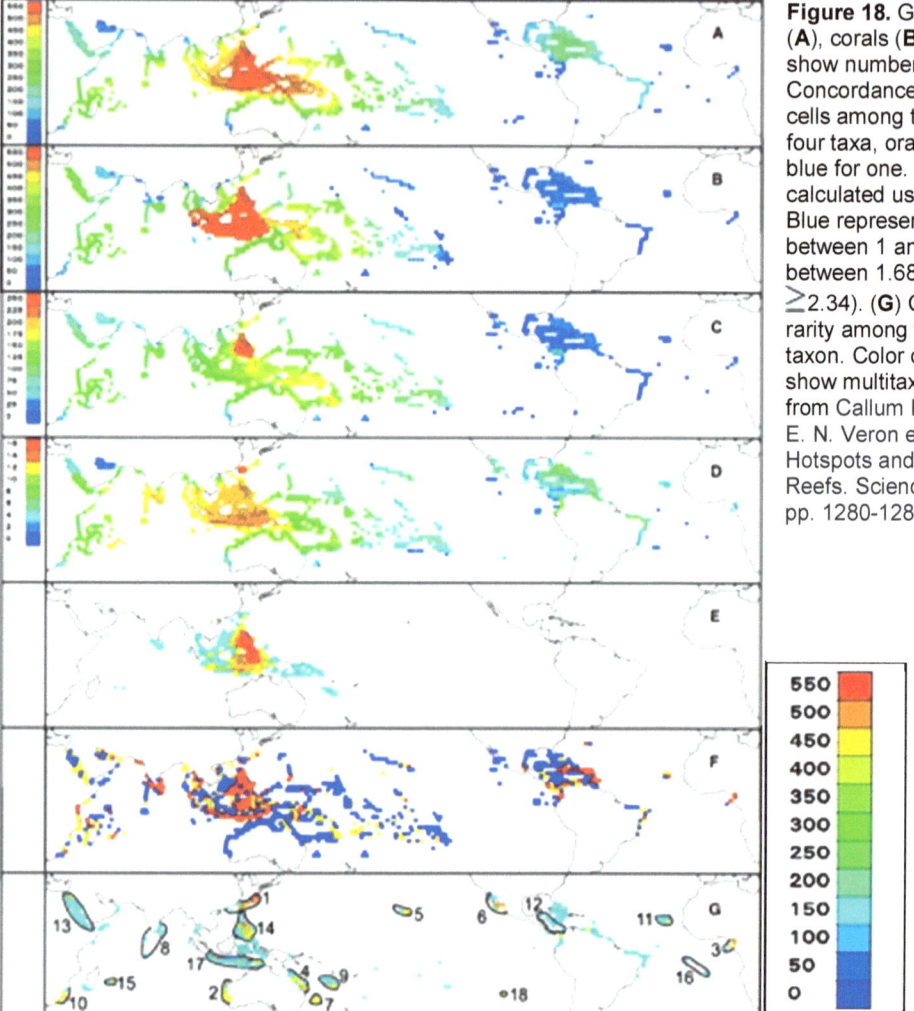

Figure 18. Global clines in species richness of fish (**A**), corals (**B**), snails (**C**), and lobsters (**D**). Scales show number of species present. (**E**) Concordance of the top 10% most species-rich cells among taxa. Red cells were included for all four taxa, orange for three, yellow for two, and blue for one. (**F**) Threats to reefs in each grid cell, calculated using data from Bryant *et al.* (1998). Blue represents low risk (ave- rage threat score between 1 and 1.67); yellow, medium risk (score between 1.68 and 2.33); and red, high risk (score ≥2.34). (**G**) Concordance in patterns of range rarity among the top-scoring 10% of cells for each taxon. Color codes are as in (E). Places outlined show multitaxon centers of endemism. (Taken from Callum M. Roberts, Colin J. McClean, John E. N. Veron et al. (2002) Marine Biodiversity Hotspots and Conservation Priorities for Tropical Reefs. Science 15th February. Vol. 295. No.5558. pp. 1280-1284).

Figure 18 E shows a high degree of overlap in the top 10% most species-rich cells for each taxon. 26.5% of the richest cells were shared by four taxa, 38.6% by three, and 38.6% by two. Cells in the southern Philippines and central Indonesia are in the top 10% richest locations for all four taxa, and degree of overlap declines moving away from this region.

Figure 18 F shows the distribution of threats to coral reefs from human impacts, based on an analysis by Bryant *et al.* (1998). Threats to reefs were mapped from coastal development, overexploitation, and pollution from marine and land-based sources, then classified reefs as facing low, medium, and high levels of threat. Using their data, we calculated the average threat to reefs in each grid cell on a scale of 1 to 3 (low to high threat)· Areas of greatest species richness are exposed to significantly greater threats from human impacts than are less rich regions.

Centres of endemism are found predominantly in isolated places such as the islands off Mauritius and La Reunion in the Indian Ocean, Hawaii and Easter Island in the Pacific and St Helena and Ascension Island in the Atlantic. They are also found where one-way ocean currents move water from tropical to temperate latitudes such as east and west Australia, eastern South Africa and southern Japan. Regions of high endemism which are highly interconnected with other regions including the Philippines, Sunda Islands and new Caledonia. The 18 richest centres of endemism can be seen in fig 19 G and include 35.2% of the world's coral reefs, covering just 0.028% of the world's oceans. These centres include between 58.6 and 68.7% of restricted range species. The top 10 richest centres for endemism cover just 15.8% of the world's coral reefs, but include between 44.8 and 54.2% of restricted range species. Figure 19 shows the threats to reefs in these centres of endemism.

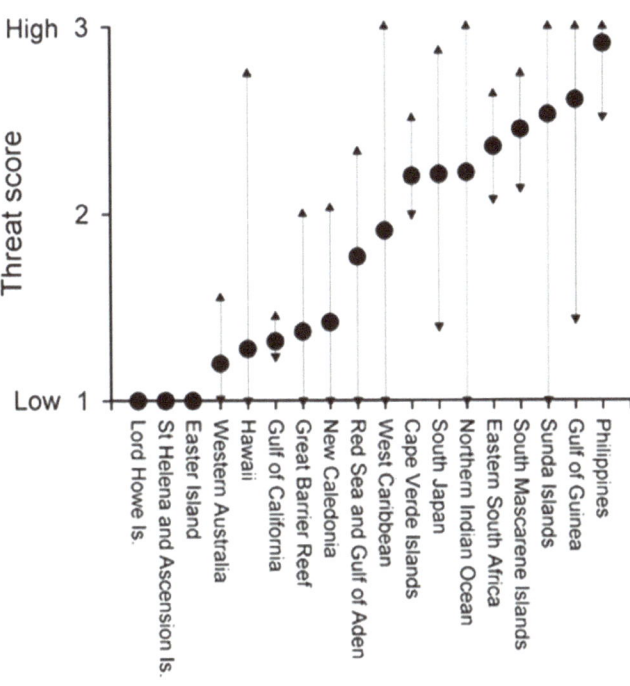

Figure 19. Threats to reefs in centres of endemism. The figure shows mean (circles), maximum, and minimum threat scores for grid cells included within each centre of endemism, calculated with data from Bryant *et al.(1998)*. (Taken from Callum M. Roberts, Colin J. McClean, John E. N. Veron et al. (2002) Marine Biodiversity Hotspots and Conservation Priorities for Tropical Reefs. Science 15th February. Vol. 295. No.5558. pp. 1280-1284).

An evolutionary perspective

Modern corals are the product of 6,000 years of growth during recent sea level rises. It is thought that overall, sea levels have risen by about 135 m over the last 10,000 years. Apart from any human induced climate warming, we are presently experiencing an interglacial with one of the warmest climates for the last 850,000 years. Sea

Fig 20. Modern corals and sponges.

Corals can only cope with increasing sea levels at a rate of approximately 4.5 cm per decade.

temperatures in the tropics have increased by almost 1°C over the last 100 yrs and rapid sea level variability, as was seen in the last interglacial, severely interrupts coral growth. Corals can only cope with increasing sea levels at a rate of approximately 4.5 cm per decade.

The rise of modern corals

Modern scleractinian corals evolved in the Triassic around 220 Ma. Their predecessors were confined to the Paleozoic and in contrast to modern corals there is no evidence that these were associated with symbiotic zooxanthellae. It may be that this symbiotic relationship began with the scleractinians when calcareous planktonic skeletons appeared and radiated in the Cretaceous. By the Cretaceous, the supercontinant Pangea had broken up and new oceans such as the Tethys were forming. The breakup of Pangea was a result of an increase in plate tectonics, which also resulted in a rise in sea levels. Continental landmasses were once again covered in warm shallow seas at low latitudes, a perfect environment for coral reef development.

~~~~RESEARCH HIGHLIGHTS~~~~

Volcanic activity recorded in coral reefs
During the Quaternary, volcanic and earthquake activity was high in the northwest area of the South China Sea. Parts of these volcanoes make up the reef substrates, while the minority of volcanoes are exposed to form reef islands. The geochemical changes associated with past volcanic activity have been recorded in the coral skeletons during their growth and it is feasible to reconstruct past volcanic activity from these coral records.
(Zhan et al. 2007)

A Cretaceous scleractinian coral

Scleractinian coral skeletons form shallow and deepwater reefs and are seen in the fossil record as far back as 240 million years ago.

Scleractinian coral skeletons are seen in the fossil record as far back as 240 mya.

A 70 million-year-coral fossil from the Upper Cretaceous, Coelosmilla sp. has been found to have skeletal structural features identical to those observed in modern scleractinians. Its skeleton, found in the Maastrichtian deposits of Poland, is completely calcitic, even though living scleractinians produce entirely aragonitic skeletons and this is thought to have been the case throughout their evolution. The calcite in this well preserved specimen has been shown to be primary, from its fine scale structure and chemistry, in other words the calcite did not form from the diagenetic alteration of aragonite as it fossilised. The aragonite found in fossil scleractinians generally dissolves or transforms into calcite during the process of fossilisation, leaving only their macroscopic morphology.

Figures 21 and 22 shows that the overall skeletal structure of *Coelosmilia* is similar to that of modern deep-sea corals, such as *Desmophyllum* and *Javania*. *Coelosmilia* sp. has a conical calice with septa arranged into five full cycles forming a hexameral pattern.

Some of the fine granulations on septal surfaces and attachment scars are still visible (fig. 21). In contrast, shells of gastropods, phragmocones of cephalopods, and other scleractinian coral species, which occur with *Coelosmilia* sp. in the same deposits, are dissolved and preserved only as molds and casts.

Fig. 21. Morphology of the modern aragonite *Desmophyllum* sp. and the Late Cretaceous calcitic *Coelosmilia* sp. (**A and B**) *Desmophyllum* sp. Relatively smooth septa, a thick septothecal wall, and a lack of pali are typical features of this solitary, azooxanthellate scleractinian coral. (**C to H**) *Coelosmilia* sp. resembles *Desmophyllum* sp. in all morphological aspects. Distal [(A), (C), (E), and (G)] and lateral [(B), (D), (F), and (H)] views are shown. Scale bar, 10 mm. (Taken from Jarosław Stolarski et al. (2007) A Cretaceous Scleractinian Coral with a Calcitic Skeleton. Science Oct 5th Vol. 318. No 5847, pp. 92-94)

The coral fossil has preserved many of the structural details, which are identical to those of living scleractinians, including radiating bundles formed by growth layers (fig 21 and 22h), vertical rods similar to trabeculae in modern scleractinians through the mid-septal zones (fig 22d), moon-shaped growth segments in the thecal reion corresponding to former growth fronts in the wall (fig 22e) and the fibre bundles indicating crystallographic axes (fig 22 h to I).

Fig. 22. Structural characteristics of the *Coelosmilia* sp. skeleton (left column) in direct comparison with the skeleton of *Desmophyllum* sp. (right column). (**A** and **H**) Optical microscopy transmitted-light images of a thin section through a septum perpendicular to the growth direction. Black arrows indicate successive growth steps in the zone of centers of calcification (COC), which appear darker than the surrounding fibrous skeleton. (**B** and **I**) Optical microscopy polarized transmitted-light images of a thin section through a septum perpendicular to the growth direction. The individual growth layers of the fibrous skeleton are clearly visible in both corals as optical layering (indicated by black arrows). (**C** and **J**) Scanning electron microscopy (SEM) images of the polished and etched surface of a septum perpendicular to the growth direction. Fiber bundles with individual growth layers are visible. (**D** and **K**) Optical microscopy transmitted-light images of a thin section through the midplane of a septum parallel to the growth direction and through the central axis of the COC. Successive COCs are organized as rods, or

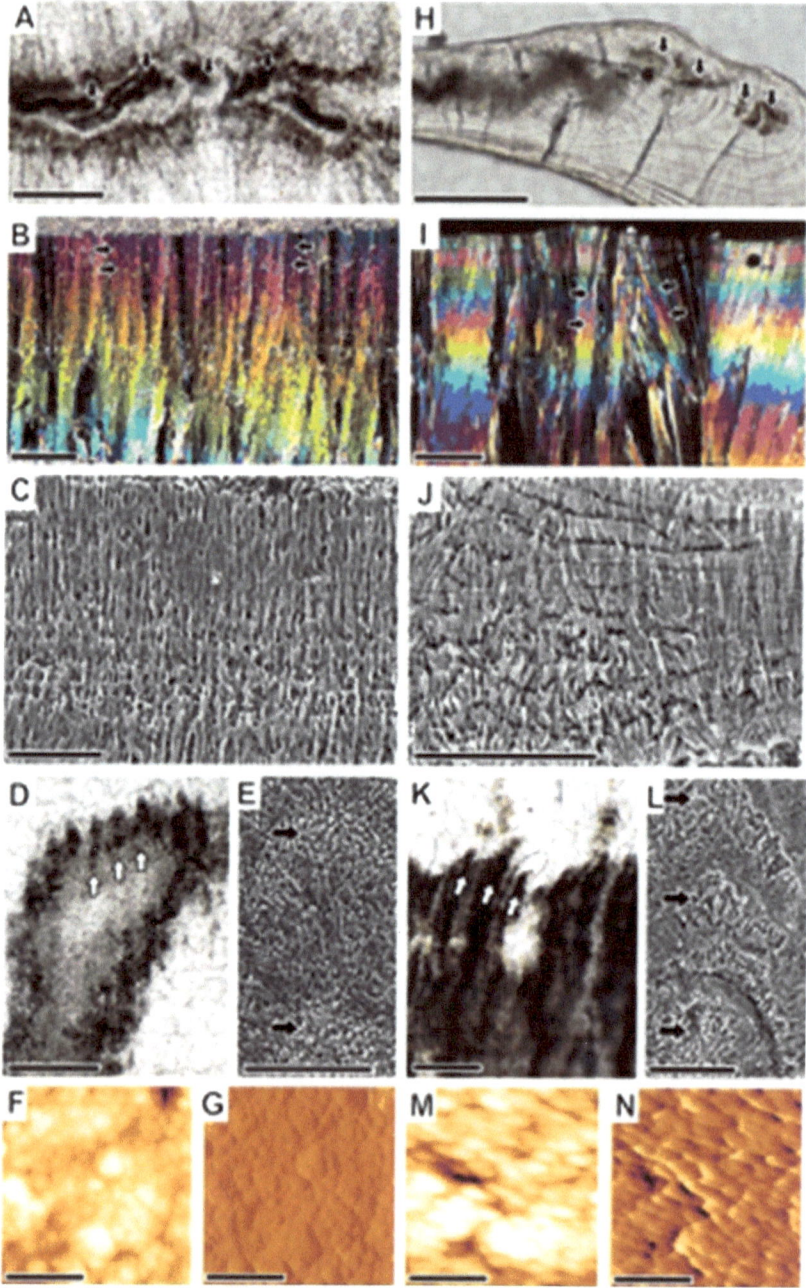

trabeculae, in the direction of growth, as indicated by white arrows. A slight undulating of the mid-septal zone trabecular centers make them appear and disappear from the plane of the cut section. (**E** and **L**) SEM images of the polished and etched surface of a cut through the thecal region parallel to the growth direction. Half-moon–shaped segments correspond to the position of organic enriched growth fronts in the wall. (**F**, **G**, **M**, and **N**) Atomic force microscopy images of skeletal fibers in height mode [(F) and (M)] and deflection mode [(G) and (N)], showing nanogranular structure. Scale bars in (A) to (E) and (H) to (L), 100 µm; in (F), (G), (M), and (N), 200 nm. (Taken from Jaroslaw Stolarski et al. (2007) A Cretaceous Scleractinian Coral with a Calcitic Skeleton. Science Oct 5th Vol. 318. No 5847, pp. 92-94).

Figure 23 shows further evidence from radiation diffraction studies demonstrating that the *Coelosmilla* fossil skeleton is made up of calcite with no trace of aragonite. This is indicated by the smaller crystallite sizes (shown as broader Bragg peaks in fig

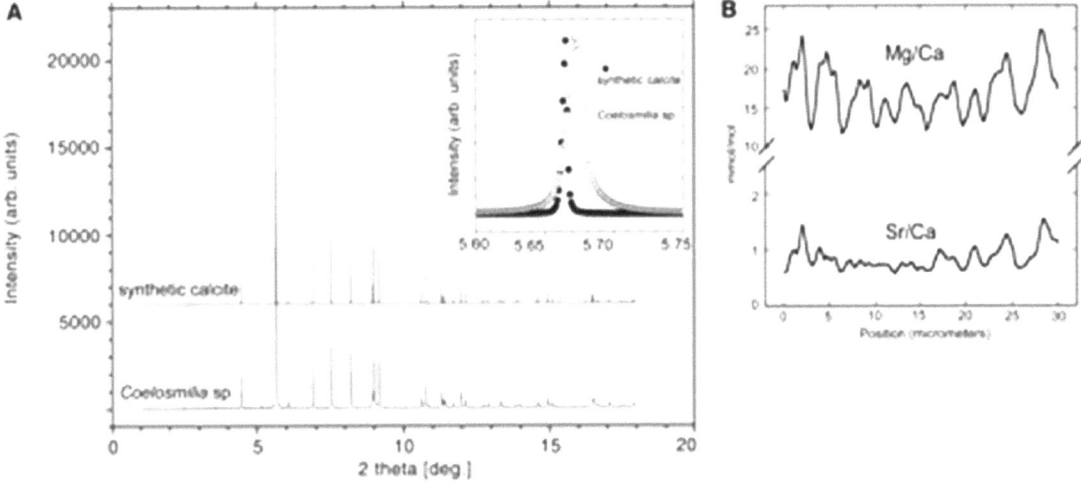

X.

Fig. 23. Crystal structure and chemical composition of the *Coelosmilia* sp. calcitic skeleton. (**A**) High-resolution synchrotron radiation powder diffraction patterns of *Coelosmilia* sp. skeleton (lower trace) and synthetic calcite (upper trace). The most intense (104) Bragg peak is shown in the inset. The Bragg peaks of *Coelosmilia* sp. are broader than those of synthetic calcite. (**B**) Typical NanoSIMS ion microprobe transect across the layered fibrous part of the *Coelosmilia* sp. skeleton. The observed variations in Mg/Ca and Sr/Ca ratios are similar in wavelength and amplitude to those observed in modern scleractinians. Taken from Jarosław Stolarski et al. (2007) A Cretaceous Scleractinian Coral with a Calcitic Skeleton. Science Oct 5th Vol. 318. No 5847, pp. 92-94

An analysis of trace elements in the fossil also indicate that it is made of calcite (fig 23 b). Overall, the structure of the *Coelosmilia* skeleton is that of a calcitic scleractinian and the calcite present has not been due to a transformation of aragonite during fossilisation, which involves a complete reorganisation of the ultrastructure of the skeleton.

Throughout geological history, the Mg/Ca ratio of seawater has changed and it has been suggested that hypercalcifying organisms, including corals, are sensitive to this delicate balance. It is thought that aragonite producing organisms such as corals flourish during periods when seawater has a Mg/Ca ratio greater than 2, and calcite producing organisms thrive when the Mg/Ca ratio is less than 2. In the modern ocean, the Mg/Ca ratio is 5.2.

The Mg/Ca ratio was below 2 when *Coelosmilia* sp. flourished in the Late Cretaceous. It has been suggested that the composition of seawater can influence the skeletal mineralogy of scleractinian corals, although aragonite producing scleractinians were present at the same time as *Coelosmilia* sp. and so skeletal mineralogy may be genetically controlled. The scleractinian corals appear in the fossil record 12 to 14 million years after the Permian mass extinctions, before which lived a weakened population of rugose corals that produced calcite. It has been suggested that increased atmospheric CO_2 levels may increase the acidity of

> Reefs could survive the large-scale environmental change that is predicted for the next century, even though they will cause major changes to the structure and function of coral reef ecosystems as we know them today.

seawater and the decalcification of carbonate skeletons, leaving naked coral skeletons (anemone-like ancestors) that survive and resume calcification when CO_2 levels decrease, thus providing a physiological refugia that ensures the survival of coral during changes in ocean chemistry.

The results of recent experiments show that corals grown in acidified conditions were able to sustain life functions and reproductive ability in a sea anemone-like form. They resumed skeleton building once re-introduced to modern marine conditions.

Thirty coral fragments from five coral colonies of the Mediterranean scleractinian species *Oculina patagonica* (encrusting) (fig 24a) and *Madracis pharencis* (bulbous) were subjected to pH values of 7.3 to 7.6 and 8.0 to 8.3 for 12 months. The corals were maintained in an indoor flow-through system under ambient Mediterranean seawater temperatures (17° to 30°C). After 1 month in these acidic conditions, morphological changes were seen, initially polyp elongation (fig 24b) followed by dissociation of the colony form and complete skeleton dissolution. The polyps remained attached to the hard rocky substrate (fig 24c).

Fig. 24. Photographs of *O. patagonica*. Scale bars indicate 2 mm. (A) Control colony. (B) Sea anemone–like coral polyps following skeleton dissolution in low-pH conditions. (C) Solitary polyps reforming a colony and calcifying after being transferred back to normal seawater following 12 months as soft-bodied polyps in low-pH conditions. (D) Time series illustrating percent change (average ± SE) in protein per polyp (biomass) and total buoyant weight over 12 months in experimental (pH = 7.4) and control (pH = 8.2) seawater (N = 20). A two-way analysis of variance (time x pH) revealed significant changes (P < 0.001) between treatments over time. (Taken from **M. Fine, D. Tchernov**, (2007) **Science** 315, 1811).

Except for just six fragments (10%), that partially lost their sybmionts (bleached) then recovered within two months, the corals maintained their algal symbionts during the experiment and all survived. After 12 months, the fragments were returned to normal ambient pH conditions and the soft, anemone-like polyps calcified and reformed colonies (fig 24d). This gives a modicum of hope that coral reefs could survive the large-scale environmental change that is predicted for the next century, even though they will cause major changes to the structure and function of coral reef ecosystems as we know them today.

It also supports the idea that the scleractinian corals which arose in the Triassic formed a robust and diverse group that indicate they originated before the Permian mass extinctions and strengthens the case for a longer evolutionary origin for scleractinian corals.

The end of the Cretaceous

By the end of the Cretaceous, 70% of corals had become extinct in one of the big five mass extinctions (see chapter three). Families of animals and plants have been shown to increase following mass extinctions in a series of radiations although there are limits to how quickly global biodiversity can recover. While 10 families of coral survived from this period, including *Acropora, Favites, Favia, Pocillopora, Porites* and *Monastrea*, 185 genera have evolved recently over the last 8 Ma. When the Tethys Ocean closed 16 Ma, relict species survived, spreading by drifting and rafting in the ocean, for example, *Pocillopora*.

Modern reefs

Today's reefs show a great diversity of species since their last period of extinction at the end of the Cretaceous, with representatives from all phyla and classes found. There are about 1,000 species worldwide of hermatypic, or reef forming corals. The Indo West Pacific being the centre of diversity with over 70 genera and around 500 species, 400 in the Philippines and around 20 genera and over 60 species in the Atlantic. As we move away from the centre of diversity the number of species decreases. For example, Midway Island in the Pacific, part of the Hawaiian chain, has just 70 species. Genera such as *Acropora* has 200 species in the Indo West Pacific but only 3 in the Atlantic and *Pocillopora* is found in the Pacific but not in the Atlantic.

~~~~RESEARCH HIGHLIGHTS~~~~

Marine biodiversity hotspots and conservation priorities
Roberts and his colleagues reported that analysis of the geographical ranges of 3,235 species of reef fish, corals, snails and lobsters revealed that between 7.2% and 53.6% of each taxon have highly restricted ranges, rendering them vulnerable to extinction. Many of these species are found in clusters within areas where reefs are affected by human activities. Ten of the riches clusters, called centres of endemism, cover 15.8% of the world's coral reefs (0.012% of the oceans) but include between 44.8 and 54.2% of the restricted range species. Conservation efforts could help maintain the valuable biodiversity of these areas.
(Roberts et al. 2002)

Reef diversity over 28 million years
More recently, however, researchers have found that although reef diversity is still poorly understood, coral diversity and reef development in a Caribbean reef system appears to be unrelated. Using fossil and stratigraphic data, they found that differences in diversity may be related to regional differences and that a high diversity is not essential to the growth and persistence of coral reefs.
(Johnson et al. 2008)

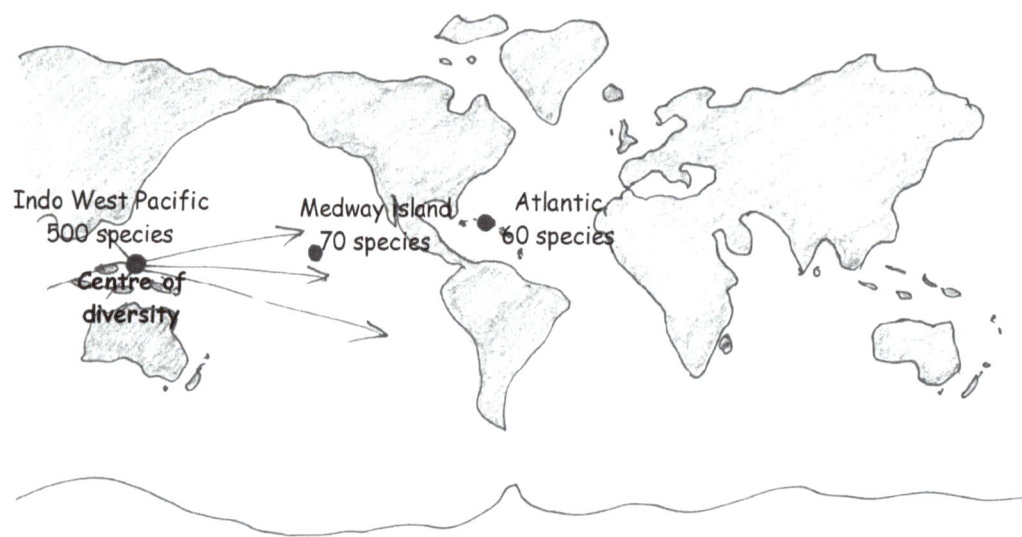

Fig 25. Coral diversity. The centre of diversity for corals is the Indo West Pacific with attenuation in all directions. The Atlantic has fewer species than the Pacific reefs, largely because of the age of the ocean and the effects of recent glaciations in the Atlantic.

Associated species and dominant species of coral also differ between oceans. This is thought to be due to differences in the age of the oceans and their evolutionary time scales. The Atlantic is relatively younger and has been subject to greater climate changes during ice ages, the last just 10-11,000 years ago. The recolonisation of this area is still in progress and Atlantic reefs usually rest on shallow banks or platforms because of wave action and erosion during the Pleistocene. Some Pacific atolls, in contrast, are around 60 million years old and show a greater complexity and diversity.

Controls on the evolution of coral reefs

~~~~RESEARCH HIGHLIGHTS~~~~

Geologic and biologic controls on the evolution of reefs

In the complex tropical reef environment there is intense competition for space, limited nutrition concentrations and a proliferation of colonial animals. Ecologically, there is a greater turnover of reef taxa than non reef taxa and this is impossible to explain simply by physiochemical changes. Biological factors which affect spatial competition are probably more important than geologic controls on reef evolution, even though climate and sea level changes undoubtedly influence the biotic composition of reefs.

Keissling W (2009)

Even though long-term climate change, sea level changes and chemical changes in the oceans influence the biotic composition of reefs, the construction of reefs during the Phanerozoic cannot be explained simply by these parameters. It seems that the biological factors affecting spatial competition may be even more important than geologic controls on the evolution of reefs (see research highlights).

Geologists and biologists have traced environmental change by studying tropical coral reefs, believing that they are sensitive to physiochemical changes in their environment. Researchers believed that reef development and even reef evolution was largely driven by these factors, but in recent years, biologic factors are also being considered as drivers of reef evolution. Tropical reefs systems are incredibly complex in terms of biodiversity, habitat structure and trophic and symbiotic interactions and these factors cannot be ignored when considering the evolution and adaptation of reefs over geologic timescales.

Evolution acts on the heritability of individuals and traditionally, researchers have believed that reef communities are simply chance associations of species with similar ecological requirements. However, it has been suggested that reefs can evolve at the community level, not as one superorganism but as collections of individual organisms that can survive disruption. This was seen in the Pleistocene when communities of coral reefs maintained their structure and stability following a drop in sea level induced by glaciations. Coral reefs show community evolution when they are ecologically stable over long periods of time.

Coral reefs show community evolution when they are ecologically stable over long periods of time.

Evolutionary turnover is greater in reefs, when compared to other ecological systems, because of their particularly complex ecology. There is greater competition for space and nutrients and because corals are colonial animals, which reduce genetic variability, they are more susceptible to long-term physiological stress and disease. Evolution should be considered a factor in explanations of reef diversity; indeed the tropics are often considered cradles of evolution with reefs producing large numbers of new species and sibling species in reefs as well as a much greater diversification in associated reef fishes.

Reef types

Some types of reef community appear to persist longer than others and the evolutionary history of reef communities shows there has been cyclical change in the type of reef community present. There have been times when reefs have proliferated and times when reefs have declined, and each of these periods seems to feature certain dominant types of coral communities. Today, there are only a few types of reef community which dominate the global reef system but in the past there have been times when the types of coral reef community have been more evenly distributed. When the numbers of reef sites within particular intervals of time are plotted against the evenness of community types present on these sites, there is a significant inverse correlation (figure 26). This implies that the expansion of coral reefs is associated with a particular type of reef community; in other words, there is a dominant type of reef community during times when reefs proliferate but a more even type of community when reefs decline.

There is a dominant type of reef community during times when reefs proliferate but a more even type of community when reefs decline.

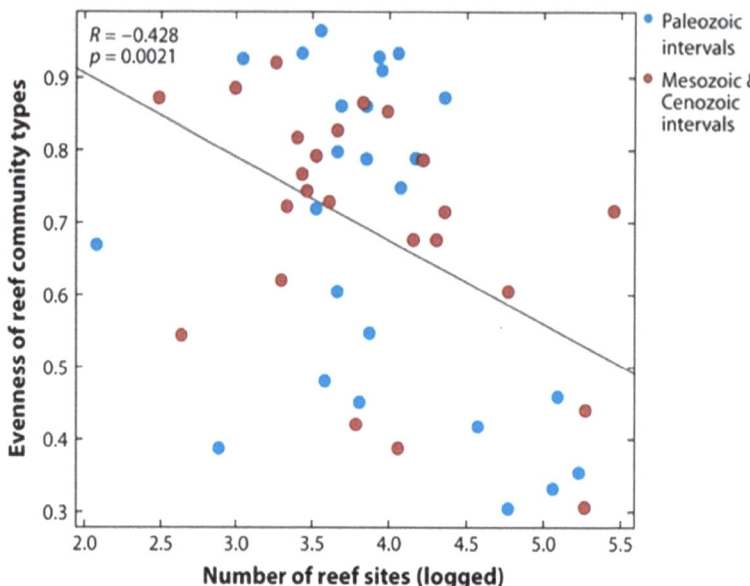

Fig 26. Number of reef sites in pre-Holocene time bins of approximately 10-million-years (Myr) duration plotted against the evenness of reef community types in the PaleoReef Database (PARED). Raw patterns are reported because reef numbers are not serially correlated. The inverse correlation suggests that only a few community types take part in reef booms, whereas community types are more varied during times of low reef numbers. Blue, Paleozoic bins; red, Mesozoic and Cenozoic bins. (Taken from Keissling W (2009) Geologic and Biologic Controls on the Evolution of Reefs. *Annual Review of Ecology, Evolution, and Systematics Vol. 40: 173-192*)

Patterns during the Phanerozoic

Booms in reef proliferation

Traditionally, anecdotal assessments show a simple and gradual increase in reef productivity until the reef collapses following a crisis, such as coral bleaching. Figure 27 illustrates the fluctuations in the number of reef sites that were recorded from each interval of time (bin) and shows how fluctuations in Phanerozoic reefs can be assessed. The pattern in figure 27 suggests that there is extreme variability in reef proliferation with short lived booms. A pattern emerges whereby reefs seem to be rare for most of the Phanerozoic, with massive and short lived booms in reef proliferation (lasting a maximum of 20 Myr). This pattern, seen in figure 27, shows fluctuations in reef proliferation with no distinct trend.

> Reefs seem to be rare for most of the Phanerozoic, with massive and short lived booms in reef proliferation (lasting a maximum of 20 Myr).

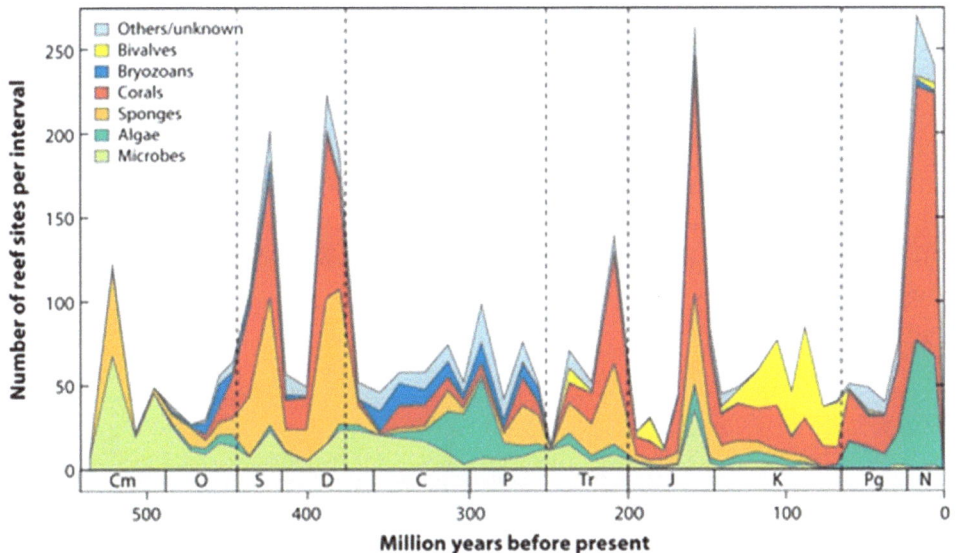

Fig 27. Number of reef sites and their biotic composition plotted by 10-Myr bin. A reef site often lumps several reef structures of the same age and environment within 20 km. The peaks persist if reef abundance is standardized for sampling, albeit with slightly different magnitudes. Vertical dashed lines indicate mass extinction episodes. Abbreviation of geological periods: Cm, Cambrian; O, Ordovician; S, Silurian; D, Devonian; C, Carboniferous; P, Permian; Tr, Triassic; J, Jurassic; K, Cretaceous; Pg, Paleogene; N, Neogene. (Taken from Keissling W (2009) Geologic and Biologic Controls on the Evolution of Reefs. *Annual Review of Ecology, Evolution, and Systematics Vol. 40: 173-192*)

Reef diversity

Also, there is no distinct trend in reef diversity, measured as the average species richness of reef building corals within a site. As figure 28 shows, reef diversity tends to develop over tens of millions of years and then, usually following a mass extinction event, diversity drops suddenly.

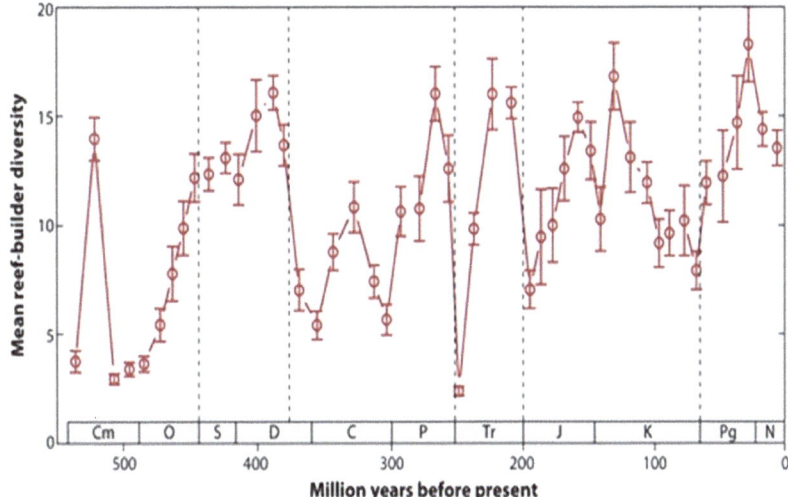

Fig 28. Trajectory of the mean species richness of reef builders in Phanerozoic reefs in 10-Myr bins (time intervals). The pattern is mostly driven by the relative abundance of low-diversity reefs. For example, the decline in the Neogene is due to abundant low-diversity coral reefs in the Mediterranean region. The trajectory differs slightly from the one reported previously (Kiessling 2005), because the definition of bins is different. Error bars represent standard errors (taken from Kiessling W (2009) Geologic and Biologic Controls on the Evolution of Reefs. *Annual Review of Ecology, Evolution, and Systematics Vol. 40: 173-192*)

Geological controls on reef development

When studying important factors that control modern reef development, researchers consider the main controls to be temperature, the availability of nutrients, sea level change and more recently, ocean chemistry. In the geological literature, sea level change has been the subject of most research into ancient reef development and evolution. However, while it is notoriously difficult to estimate many of the other physicochemical parameters as they relate to the reefs of the past, a review of the literature reveals some interesting findings.

Climate change

The effects of climate change on modern reefs is the subject of much research and concern, but over a vast geological timescales, changes in the mean global temperature of the past, reconstructed from stable oxygen isotopes, do not correspond to changes in either reef abundance or latitudinal distribution of reefs from the Phanerozoic. While some studies have linked the expansion and decline of reefs with climate

> Some studies have linked the expansion and decline of reefs with climate change, there are big holes in the argument, since the relationship between coral reefs and climate change is not a simple linear relationship.

change, there are big holes in the argument, since the relationship between coral reefs and climate change is not a simple linear relationship. However, reefs do trace the palaeoclimate indirectly with different reef communities prevailing during various climatic phases. Algal reefs tend to be dominant during icehouse conditions such as during the late Carboniferous-Early Permian Gondwana glaciation, when ice moved towards tropical latitudes. Rather than temperature per se, ocean circulation patterns that brought excess nutrients may have allowed algal communities to dominate and ocean chemistry and other factors may have also played an important role in the type of reef communities that were present.

During greenhouse conditions, a well-defined reef zone formed in the tropics with a concentration of diverse reef types in different latitudes. Coralline sponge and microbial reefs dominated the lower latitudes with coral reefs abundant in intermediate latitudes. The decline of coral reefs in the late Cretaceous and the spread of rudist reefs were partly due to global warming, since rudists had a greater tolerance for high temperatures than corals. However there are other factors which play a part in the success and decline of corals that may be linked to reef evolution.

Ocean chemistry

Ocean acidification is just as critical to coral reefs as ocean temperature and has received much attention by scientists in recent years. Reef growth is closely linked to organisms that precipitate the calcium carbonate that forms the reef, such as calcareous algae, corals, sponges, rudists and bryozoans. There is a good correlation between the abundance of these hypercalcifying taxa on calcium

> Ocean acidification is just as critical to coral reefs as ocean temperature.

carbonate ($CaCo_3$) substrates, recorded in the Palaeobiology Database, and periods of high reef growth. This relationship is seen by comparing figure 28 (reef growth in the Phanerozoic) with figure 29 (number of reef building taxa).

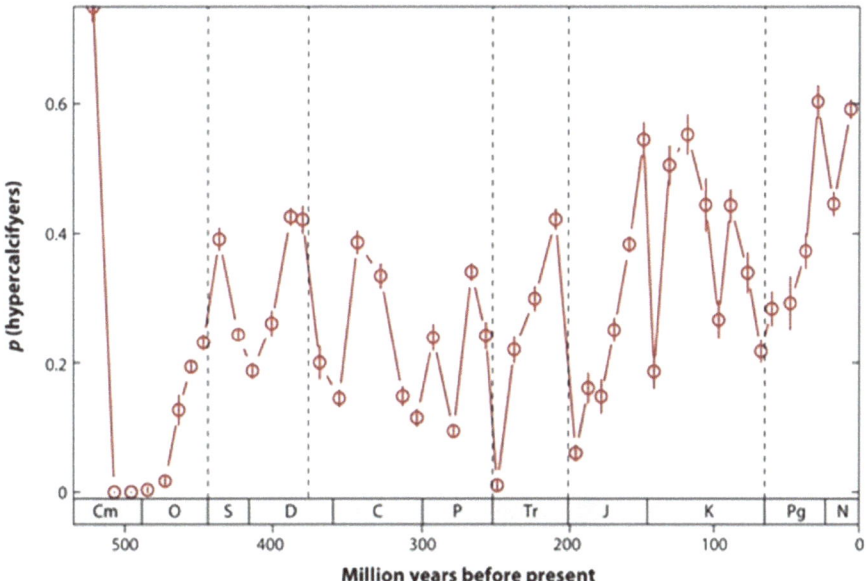

Fig 29. Number of hypercalcifying taxonomic occurrences (calcareous algae, corals, sponges, rudists, bryozoans) relative to all epifaunal macrobenthic occurrences on calcium carbonate substrates recorded in the Paleobiology Database. Error bars represent standard errors. (Taken from Kiessling W (2009) Geologic and Biologic Controls on the Evolution of Reefs. *Annual Review of Ecology, Evolution, and Systematics Vol. 40: 173-192)*

However, one thing that puzzles researchers is that various no significant cross correlation between changes in atmospheric pCO_2, which impact ocean pH, and changes in the reef attributes. The long-term changes in the saturation state of ocean water do not explain the proliferation and decline of reefs and hypercalcifying taxa. According to models, CO_2 concentrations in the atmosphere were higher for most of this history than they are today, yet despite several studies into atmospheric carbon dioxide concentrations and changes in the saturation state of the oceans during the Phanerozoic, the link with Hypercalcifying taxa or the distribution of calcium carbonate sediments as an indicator of pH remains a mystery, although acidification events during Mass Extinctions could be implicated in the decline of ancient reefs.

One study linked ocean chemistry and reef evolution with the Mg-Ca ratios in seawater, resulting from plate tectonic activity. The Mg-Ca ratios

~~~~RESEARCH HIGHLIGHTS~~~~

Could ancient reef building be driven by plate tectonics?
A study on the evolution of coral reefs suggests that there is a link between ocean chemistry and the predominant mineralogy of reef building organisms at various times through evolution. The researchers suggested that the tectonic activity controls the Mg-Ca ratios in seawater and drive the activity of reef building organisms. They found that when the Mg^{2+} concentration is low, calcite is precipitated from solution but aragonite is precipitated when the Mg-Ca ratio is greater than two. In biological terms, this means that scleractinian corals (aragonite reef builders) are prevalent during periods of high Mg-Ca, so-called aragonite seas, and calcitic reef builders are more prevalent during periods of low Mg-Ca. Although an intriguing hypothesis, it does not completely explain the increase and decline of reefs or reef diversity. Stanley SM; Hardie LA (1998).

contribute to the success of reef building organisms and may explain the dominance of certain reef builders at different times during their evolution. Experiments showed that calcite is precipitated from aqueous solution when there are low Mg^{2+} concentrations and aragonite is facilitated when the Mg-Ca ratio is greater than two. This can be translated into biological terms by models that show that scleractinian corals were more prevalent in aragonite seas, when the Mg-Ca ratio was high, and calcitic reef builders were more prevalent when the Mg-Ca ratios were low. Although this shows an alternative hypothesis for the pattern of reef diversity, it does not tell the whole story since these 'rules' appear to have been reversed during the late Cenozoic and late Jurassic periods.

Changes in sea level

Scleractinian corals were more prevalent in aragonite seas, when the Mg-Ca ratio was high, and calcitic reef builders were more prevalent when the Mg-Ca ratios were low, although these 'rules' appear to have been reversed during the late Cenozoic and late Jurassic periods.

Sea level changes are thought to trigger mass extinctions in coral reef communities and are related to the availability of shallow water habitats, suitable for coral reef development. Major changes in sea level during periods of glaciation have had a significant effect on the structure and evolutionary rates of topical marine species. Although other factors influence coral reef evolution, sea level changes must surely have profoundly affected the evolutionary pattern of reefs during the Pleistocene. Sea level changes are easily accessible factors for geological study and correlations with the abundance of reefs have been estimated by modelling over million year timescales.

Nutrient availability

The availability of nutrients in the ocean is variable over time and while climatic changes appear to co-vary with changes in the level of nutrients, it is not well understood.

From studies on the increases in biomass of invertebrate skeletons, metabolic activity, the abundance of phytoplankton and patterns of strontium, carbon and sulphur stable isotopes, it is thought that nutrient levels have increased through the Phanerozoic.

A lack of nutrients is detrimental to any living organism. Sometime during their evolutionary development, coral reefs developed photosymbiosis, a symbiotic relationship with zooxanthellae and other microorganisms, and were able to thrive in oligotrophic environments. This symbiosis reduces competition with sessile organisms, such as filter feeders, that require a continual supply of nutrients. It has been argued that the evolution of photosymbiosis was a response to competition for space and or lack of nutrients.

It has been argued that the evolution of photosymbiosis was a response to competition for space and or lack of nutrients.

Global reef distribution patterns can interfere whether this mutual symbiosis allowed reefs to thrive in areas of low nutrient availability.

38

Photosymbiosis occurred in scleractinian corals from the Triassic period and later in some rudists from the Cretaceous, however, it was not apparent in early reefs.

Continental configuration and the likely surface currents in past marine environments can show whether reefs grew close to upwelling zones, which are rich in nutrients. During a prolonged period of time form the Carboniferous to the early Permian, reefs were dominated by microbes and algae that required nutrients, and for a short period during the early Jurassic, a variety of semi-infaunal bivalves that were important reef builders also depended on nutrient rich seas. From this, we can infer that most of the Phanerozoic reefs were able to survive in nutrient depleted habitats, with reefs of the Palaeozoic more strongly adapted than their ancestors. Isotopic studies suggest that photosymbiosis occurred in scleractinian corals from the Triassic period and later in some rudists from the Cretaceous, however, it was not apparent in early reefs.

Biotic controls

The evolutionary history of coral reefs shows an increase in biological disturbance. In other words there was an increase in specialised predators, bioeroders and herbivores during the Cretaceous and Cenozoic, which influenced the community structure of reefs.

An increase in specialised predators, bioeroders and herbivores during the Cretaceous and Cenozoic influenced the community structure of reefs.

Sessile organisms such as corals require a stable substrate that is rarely disturbed or vulnerable to the accumulation of sediments. In a marine environment, these types of habitat are very limited and competition for space is fierce.

Mass extinctions

There have been five major extinction episodes: the end of the Ordovician (444 Mya), the late Devonian (375 Mya), the end of the Permian (251 Mya), the end of the Triassic (200 Mya) and the end of the Cretaceous (65 Mya). While mass extinctions can cause catastrophic changes to ecosystems generally and to the decline of reef diversity and carbonate production in tropical green environments, when we consider calcium carbonate production as an indicator of the health of coral reefs, only the first for extinction events actually qualify as reef crises. The end of the Cretaceous extinction event saw a catastrophic decline from a genealogical perspective, with declines in reef-building species and the size of the reef population, especially zooxanthellae corals, but reef diversity and carbonate production is not generally considered in the literature to be a major reef crisis.

Some researchers have suggested that ocean acidification may be the major factor responsible for reef crises, although this has been disputed and finding reliable evidence to prove the argument either way has been problematic. Following mass extinctions, some important reef building groups suffered, particularly during the late Devonian with extinctions of stromatoporoids and tabulate corals. The extreme extinctions at the end of the Permian brought profound changes in coral taxonomy, but the ecological changes in reefs appear to be fairly modest once the reef had recovered. The stony corals of the Palaeozoic became extinct but several genera of calcareous sponges survived to build reefs in the Permian and Triassic.

Disturbance to coral reefs

Previous thought was that diversity in animal and plant species was due to stability and equilibrium on coral reefs but now it is recognised that disturbance is important in opening up niches for corals to grow. Studies at Heron Island on the Great Barrier Reef show that continual moderate change in conditions increased diversity, the 'intermediate disturbance' hypothesis, although other studies have suggest that the role of disturbance is more subtle than has been believed.

> Disturbance is important in recolonisation and distributing genetic material and is thought to be an evolutionary adaptation to increase prospects for survival.

Catastrophic mortalities caused by El Nino Southern Oscillation (ENSO) events as well as tropical storms and hurricanes can open up niches which may otherwise be dominated by a few species and look very different to the rich environments we know today. The fragmentation which results from disturbance is important in recolonisation and in distributing genetic material and is thought to be an evolutionary adaptation to increase prospects for survival.

Behaviour of corals

Amazing behaviour has been studied in corals with aggression noted in many coral species. Corals attack neighbours with their tentacles and there seems to be a hierarchy of aggression and dominance which is different in the Atlantic and the Indo West Pacific. The Atlantic reefs are dominated by *Acorpora palifera* (elkhorn coral) and *Millepora complanata*, a hydrocoral. (*Millepora* does not dominate reefs in the Indo West Pacific). In the Indian Ocean, *Favia favus* and *Acropora palifera* are for example more aggressive than the subordinate *Monipora* or *Porites* species. These amazingly competitive networks are thought to encourage diversity in coral species.

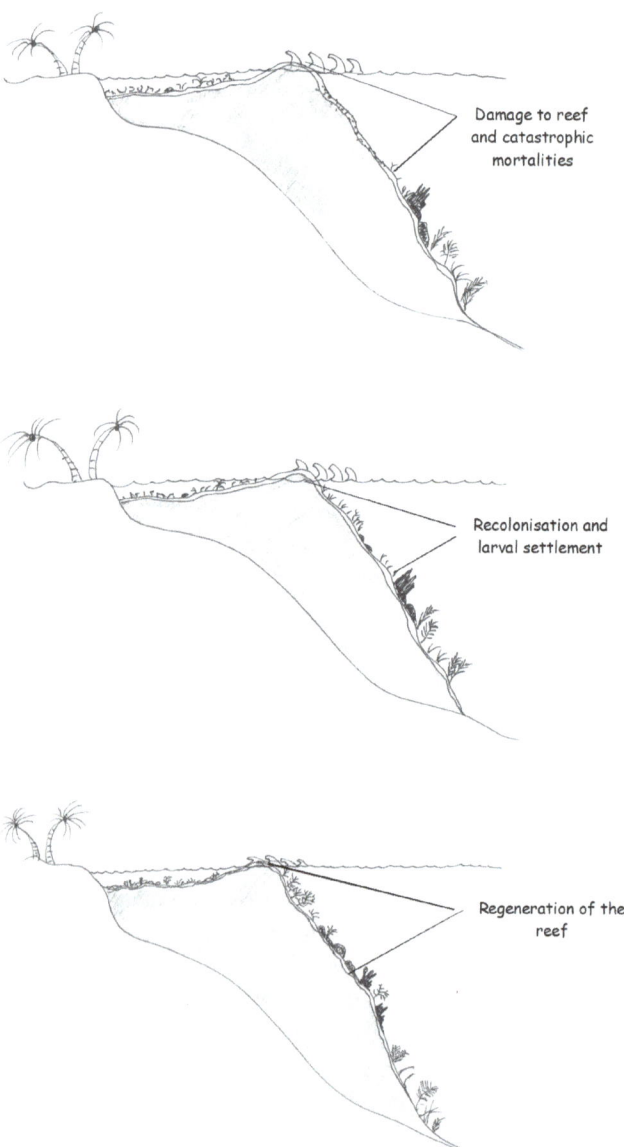

Damage to reef and catastrophic mortalities

Recolonisation and larval settlement

Regeneration of the reef

Fig 30. Catastrophic mortality. Catastrophic mortality on a reef may be important to genetic diversity.

Associated species

Associated species are also rich in diversity on reefs with around 500 species of reef fish in the Bahamas, 1,500 in the Great Barrier Reef and 2,000 in the Philippines. Fish are the major grazers in the Pacific while echinoids fill the role in the Atlantic. Mangroves and sea grasses also show high diversity in the Indo West Pacific in contrast to the Atlantic with 9 species of mangrove in the Atlantic and around 35 in the Indo West Pacific. Atlantic reefs also lack the giant clams such as *Tridacna* and the octocorals, *Helipora* and *Tubipora* and there are no anemone fish, *Amphiprion* in the Atlantic although there are fish which associate with anemones. The crown of thorns starfish, *Acanthaster* is also absent in the Atlantic.

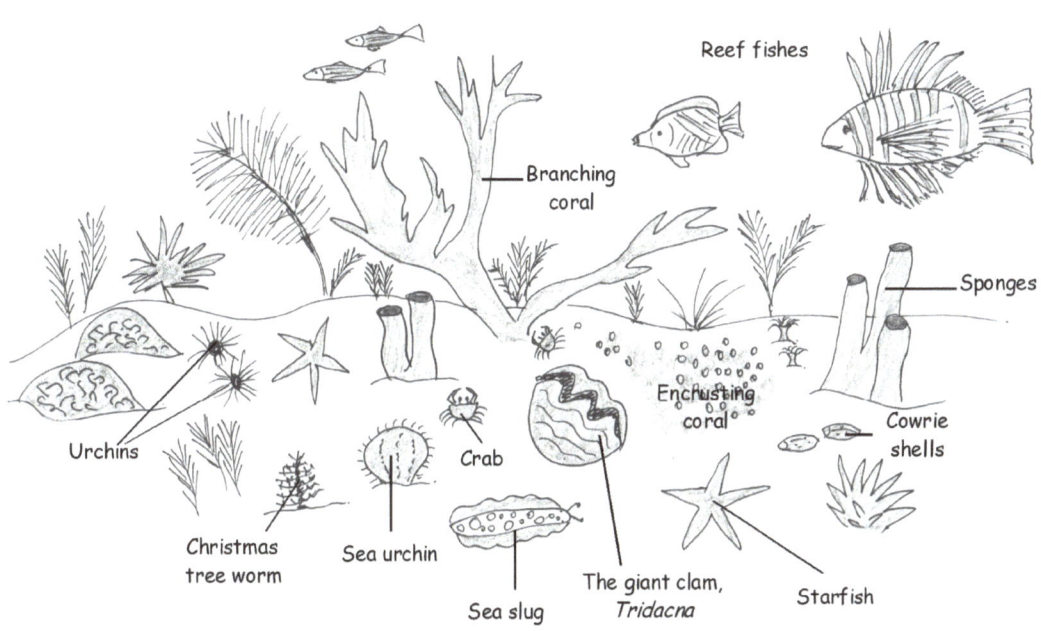

Fig 31. Associated species. Reefs are home to a great diversity of marine species.

Representatives of almost all the phyla are found within the coral reef ecosystem. The reef itself, with its nooks and crannies, is home to many plants and animals with a variety of microhabitats for small invertebrates, sessile animals and fish that hide away such as moray eels.

> Representatives of almost all the phyla are found within the coral reef ecosystem.

Much coral rubble and sand builds up between crevices and is an important home and feeding area for burrowing animals and deposit feeders.

Many bottom dwellers as well as herbivorous fish graze over the hard substrates for algae. Invertebrates include porifera (sponges); echinoderms such as starfish, urchins and sea cucumbers; molluscs such as limpets, snails and clams and arthropods such as crabs, lobsters and shrimps.

Fig 32. Coral reef animals.

Vertebrates include a vast array of brightly coloured fish and it is estimated that some 25% of the marine fishes in the world are found only in reefs. Most fish are specialized feeders; some are herbivores; some feed on plankton and others are piscivores (they feed on other fish) or carnivores, which feed on marine invertebrates. Reef fishes include snappers, *Lutjanus* sp; parrotfish, *Scarus* sp; groupers, *Epinephelus* sp; butterfly fishes, *Chaetodon* sp. And moray eels, *Gymnothorax* sp.

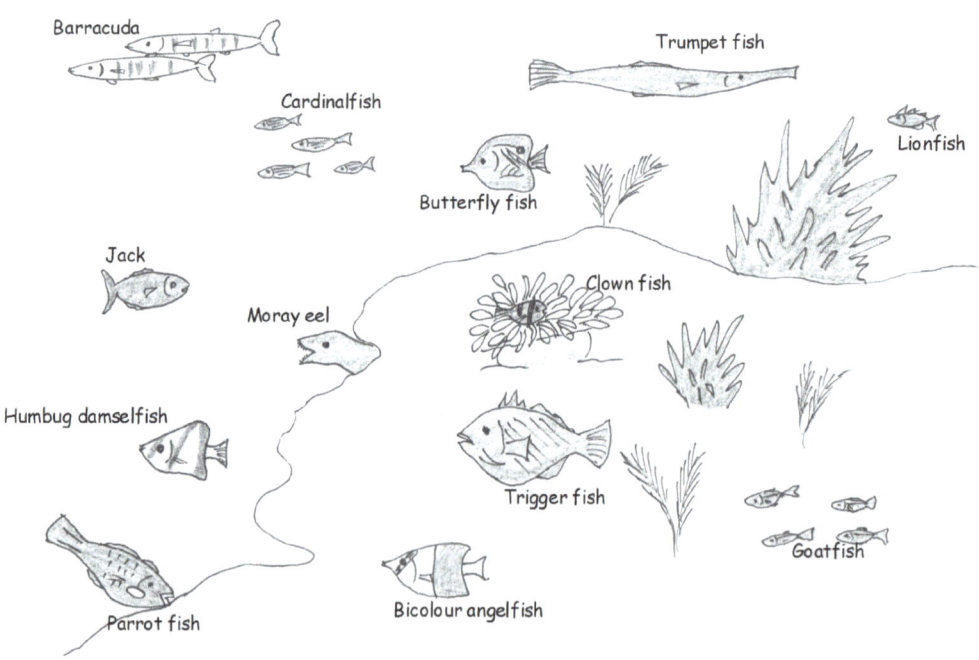

Fig 33. Reef fishes. Reef fish play an important part in the ecology of a coral reef.

Reef fish

Fish are important to the ecology of the reef and play a role in determining the zonation of the reef. By grazing algae, they keep certain species in check and prevent them dominating the corals. Some fish, such as parrotfish also graze on the corals themselves, which not only opens up niches for new corals but by excreting the ground-up calcareous remains, they contribute to the formation of sandy sediment for burrowers and deposit feeders to use as habitat. The reef is transformed at night, with many fish that feed in the day retiring to the shelter of the cracks and crevices of the reef for the night.

~~~~RESEARCH HIGHLIGHTS~~~~

Sunscreen in fish
Researchers in Hawaii found sunscreen compounds in the epithelial mucus of 138 different species of reef fish that absorb both UVB and UVA (320 to 400 nm) radiation. The concentration of these sunscreens changes according to the UV exposure of the fish (Zamzow and Losey 2002).

Another study at 3 different locations along the Great Barrier Reef showed that the sunscreen compounds found in reef fish mucus not only varies among species but also within a species sampled at different locations (Eckes et al. 2008).

Fig 34. Parrotfishes (Scaridae). Predation on coral polyps by parrotfishes can cause >50% coral mortality in some areas. See Research Highlights below.

Marine biologists are interested in how reef fish are recruited onto the reef to better understand their population dynamics and manage and conserve marine biodiversity. One aspect that has been studied is the dispersal of fish larva following spawning. It has

Some 25% of the marine fishes in the world are found only in reefs. Most fish are specialized feeders and are important to the ecology of the reef.

generally been assumed that fish larva is dispersed passively on ocean currents over large areas of the ocean and although ocean currents are important for larval dispersal, studies have shown that many fish larva are retained on the reef where they were spawned. Incredibly, fish larva appear to be able to smell their home reef and prefer to settle there than any other reef – they are able to discriminate because each reef has a distinctive smell.

~~~~RESEARCH HIGHLIGHTS~~~~

Coral predation by parrotfishes
Direct predation by parrotfishes (Scaridae) may be an important threat to for reef building corals and were shown to cause >13% mortality in the coral *Porites asteroids* in a Belizean back reef habitat. Heavily grazed colonies showed >50% tissue loss to parrotfish predation. The most heavily grazed areas included macroboring organisms such as barnacles, polychaetes and molluscs, which, it is suggested provide additional nutrition for parrotfishes and attract them preferentially to these areas.
(Rotjan et al. 2005)

Researchers first tested this idea by marking the otoliths, or ear bones, of over 10 million developing embryos of the damselfish, *Pomacentrus amboinensis*, at Lizard Island in the Great Barrier Reef. They found that when they examined 5,000 juveniles settling at the same location, 15 of them were marked individuals. They estimate that as many as 15-60% of juveniles may be returning to their natal reef, termed self-recruitment. This homing instinct is also seen in many other marine animals including nesting turtles that return to their natal beaches and even limpets that return to a home scar following a foraging expedition.

Fish diversity

The cleaner fish *Labroides dimidatus* is an important species on the reef. In fact, around 2297 client fish interact with a single cleaner fish each day on the Great Barrier Reef and a single client fish can visit a cleaner fish up to 144 times per day. That amounts to the removal of an estimated 1218 parasites per day, per cleaner fish. When cleaning does not take place, researchers found that the parasite load of *Hemingymnus melapterus* fish that were caged on a coral reef increased by 4.5-fold within 12 hours.

The cleaner fish *Labroides dimidatus* is an important species on the reef.

It has long been thought that cleaner fish are attracted to areas where client fish are present and respond to patterns of client distribution. However, one 18 month study covering some 18 separate reefs has shown that the cleaner fish, *L. dimidatus* may itself have a disproportionate effect on the diversity and activity of many other coral reef fish species. In an experiment where *L. dimidatus* was excluded from small reefs for 18 months, the local fish diversity was significantly affected. Species diversity was reduced by 50% and there was just one-fourth the abundance of individuals. The fish affected were large species that move among other reefs, rather than the resident species and they also tend to have an effect on other reef organisms.

44

It seems that one small cleaner fish can influence the movement patterns, activity, habitat choice, local diversity and abundance of many other reef fish species and that reef fish appeared to choose reefs where cleaner fish are present.

Figure 35 shows that the number of species of visiting client fish as well as the number of visiting individuals, when sampled by remote video and by a snorkeler, were two- and four-fold higher, respectively, on reefs with cleaner fish than on reefs without.

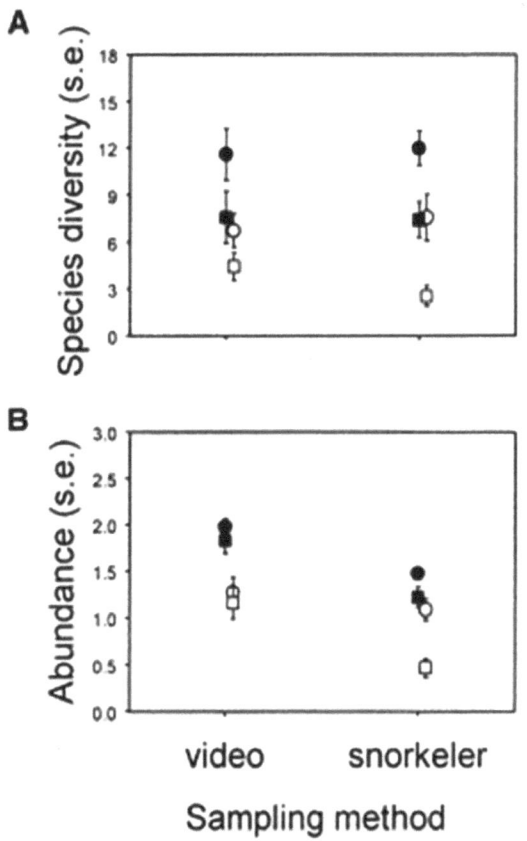

Figure 35. Mean (s.e.) Visiting client fish numbers per reef at two sites counted by remote video and by a snorkeler

(A) The number of visiting species on reefs with (closed symbols) and without (open symbols) cleaner fish at Casuarina Beach (circles) and Lagoon (squares) sites.
(B) The $\log_{10}(x + 1)$ abundance of visiting client fish per reef. Symbols are as in (A). Counts in (A) and (B) were pooled across different times of the day. (Taken from Alexandra S Grutte, Jan Maree Murphy and J.Howard Choat (2003) Cleaner Fish Drives Local Fish Diversity on Coral Reefs. *Current Biology*. Vol 13, Issue 1. P64-76).

When sampled by a scuba diver, figure 36 shows that reefs with cleaner fish also had more species of visiting clients than reefs without cleaner fish, but the difference was less. This implies that visiting clients were disturbed by scuba divers but not by the remote video or snorkelers.

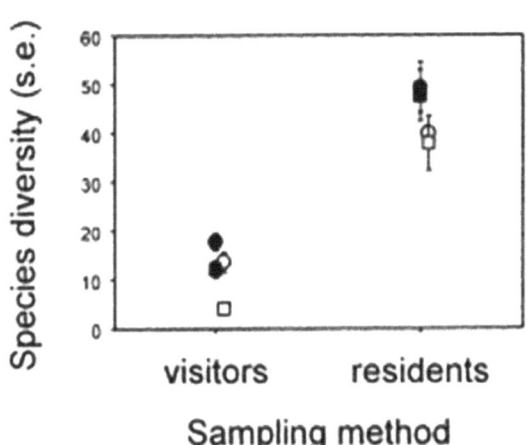

Figure 36. Mean (s.e.) Visiting and resident client fish species per reef at two sites as counted by a scuba diver. (Taken from Alexandra S Grutte, Jan Maree Murphy and J.Howard Choat (2003) Cleaner Fish Drives Local Fish Diversity on Coral Reefs. *Current Biology*. Vol 13, Issue 1. P64-76).

Table 1 shows that in total, there were 78 visiting species on these reefs, 38% were found only on reefs with cleaner fish, 52% and 9% were only found on reefs without cleaner fish.

% of visiting species	Family
38% found only on reefs with cleaner fish	Acanthuridae, Balistidae, Carangidae, Dasyatidae, Haemulidae, Holocentridae, Labridae, Lethrinidae, Lutjanidae, Priacanthidae, Scaridae, Siganidae, and Sphyraenidae
52% found both on reefs with and on reefs without cleaner fish	Acanthuridae, Ephippidae, Hemigaleidae, Haemulidae, Holocentridae, Labridae, Lethrinidae, Lutjanidae, Mullidae, Nemipteridae, Pomacanthidae, Scaridae, Siganidae
9% only found on reefs without cleaner fish	Acanthuridae, Carangidae, Carcharhinidae, Holocentridae, Labridae

Table 1. The percentage of visiting species on the reefs relative to cleaner fish presence.

While other studies on the same reefs have not found that cleaner fish effect either the number of fish species or their abundance, they did not separate fish into residents and visitors. This study looked at the differences between resident fish and visitors between reefs and found that the biggest effect was seen in visiting fish. This could be because resident fish are generally small and may have fewer parasites or that the costs of seeking cleaners, such as increased predation risk when moving between reefs, loss of territory and the energy costs, outweigh the costs of not being cleaned. However, smaller fish may be more vulnerable to the parasite than larger fish.

> Cleaner fish may be important to reef health and diversity and may play a key ecological role.

Introducing cleaner fish to artificial or damaged reefs may help to improve fish diversity, and the practice of removing cleaner fish for the aquarium trade should be strongly discouraged. Cleaner fish may be important to reef health and diversity and may play a key ecological role.

~~~~RESEARCH HIGHLIGHTS~~~~

Barcodes for fish

The trade in ornamental fish needs to be regulated as part of the conservation management of tropical ecosystems. Identification of species is therefore an important, yet difficult task. Fish ID can be made more accessible by assembling a DNA barcode reference sequence library. ID of 391 species from 8 coral reef locations was possible as 98% of these species showed barcode clusters in their genes.
(Steinke et al. 2009)

Fig 37. Barcodes for fish. Aquarium fish could be catalogued using their DNA.

Moray eels are thick bodied fish with no pectoral or pelvic fins. They ambush prey from reef crevices with their powerful jaws. There are over 100 species.

Fig 38. The moray eel. The Moray eel is a key predator on the reef.

~~~~RESEARCH HIGHLIGHTS~~~~

The fearsome jaws of moray eels

Most bony fishes rely on suction to capture prey into their mouths but this mechanism is less effective in the moray eel, *Muraena retifera*, and it is not known how moray eels swallow large fish and cephalopods. This study shows that moray eels overcome the problem of reduced suction by launching raptorial pharyngeal jaws from its throat and into its oral cavity, where the jaws grasp struggling prey and transport it back to the throat and into the oesophagus. This second set of pharyngeal jaws is extremely mobile, thanks to elongated muscles that control the jaws and allow them to reach up from behind the skull, similar to the ratcheting mechanisms used in terrestrial snakes.
(Mehta et al. 2007)

We met sharks in Chapter five with our overview of vertebrates, and again in chapter seven when we discussed marine fishes in more detail. As a reminder, sharks belong to the class Condrichtheys, the cartilaginous fishes, and sharks and rays are grouped into a subclass, Elasmobranchii. Collectively, cartilaginous fishes make up around 5% of the world's fishes, with about 330 species of shark and 450 species of skates and rays.

Cartilaginous fishes make up around 5% of the world's

Fig 39. A lemon shark. A Lemon shark, *Negaprion brevirostris* in the Bahamas.

Sharks are most abundant in the warm temperate and tropical waters of the world. The females of many viviparous species give birth to live young in special nursery areas, while oviparous species lay their eggs in coastal bays, coral reefs, atolls and mangroves, where they are relatively safe from larger sharks. Oviparity and viviparity in sharks is discussed in chapter five.

Figure 39 shows a lemon shark, *Negaprion brevirostris*. These viviparous sharks give birth to around 8-12 live pups. They are found mostly in the Caribbean, around reef systems with areas of associated seagrass and mangroves, which they use as nursery areas. Other viviparous reef sharks, include grey reef sharks, *Carcharhinus amblyrhynchos*, whitetip reef sharks, *Triaenodon obesus*, blacktip reef sharks, *Carcharhinus melanopterus,* great hammerheads, *Sphyrna mokarran* and scalloped hammerheads, *Sphyrna lewini* .

Hammerhead sharks are found in shallow tropical waters worldwide. Scalloped hammerheads, shown in figure 26, are known for their schooling behavior around seamounts, where hundreds of individuals, predominantly females, aggregate during the daytime. At night, like other sharks, they are solitary hunters. There is still considerable debate about why

Fig 40. The scalloped hammerhead shark, *Sphyrna lewini*. Scalloped hammerhead females aggregate in huge numbers around seamounts during the day.

48

scalloped hammerheads school in such huge numbers, but recent suggestions include socializing and using seamounts as a navigational marker, related to the Earth's magnetic field. The great hammerhead, *Sphyrna mokarran*, favours offshore and inshore reefs, coral drop-offs and adjacent sandy habitats.

The most common reef sharks in the Pacific and Indian oceans include whitetip reef shark and blacktip reef sharks. Both species prefer shallow lagoons and coral reefs close to shore and tend to stay around the reef. Caribbean reef sharks, *Carcharhinus perezi,* are found most commonly around the island reefs of the tropical and sub-tropical waters of the Caribbean and the Gulf of Mexico. They can also be found near coral drop-offs on the outer edges of the reef down to about 30 m depth.

Fig 41. The blacktip reef shark, *Carcharhinus melanopterus.* The blacktip is a common shark around reefs in the Pacific and Indian oceans.

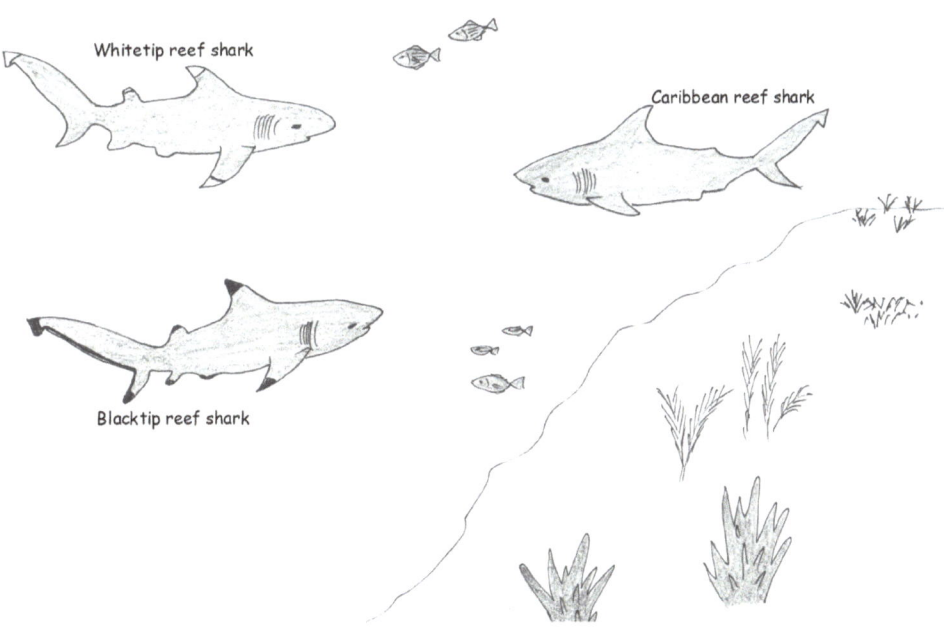

Fig 42. Reef sharks. Several species of shark visit the reef, including whitetip and blacktip reef sharks in the Pacific and Indian oceans. Caribbean reef sharks are found in in the western Atlantic.

Visitors to the reef include tiger sharks, *Galeocerdo cuvier*. They are the only ovoviviparous members of the family of requiem sharks. Females carry between 6 and 82 pups and the developing embryos may eat their smaller siblings, a practice known as inter-uterine cannibalism. Sand tiger shark pups, *Carcharias taurus*, also practice inter-uterine cannibalism and only 2 pups are born at the end of gestation – one from each horn of the uterus.

Fig 43. The tiger shark, *Galeocerdo cuvier*. Tiger sharks are the most common large shark found in tropical waters worldwide.

The largest fish in the sea, reaching sizes of up to an estimated 18 m, is the whale shark, *Rhincodon typus*. Whale sharks are found in warm temperate and tropical waters worldwide and are occasional visitors to lagoons and coral atolls. Despite their enormous size, they feed mainly on zooplankton and also eat small fish and squid.

Fig 44. The whale shark, *Rhincodon typus*. The whale shark may occasionally enter lagoons and coral atolls.

Sharks face number of threats, including deliberate and incidental capture by fisheries. Fisheries and sports fishermen target 'trophy sharks' and there is a considerable demand for the jaws and teeth of many shark species. Shark finning is also a huge threat, a trade that targets sharks for their fins. The fins are so valuable that once the fins have been removed, the rest of the shark is thrown back into the sea – often still alive. This barbaric practice, which has killed over 28 million sharks, continues because of the huge market demand for shark fin soup in countries such as China, Hong Kong and Taiwan. The pressure to catch sharks has spread around the oceans of the world, wherever sharks are found, including coral reefs (we will cover these and other threats to sharks in more detail in chapter eighteen).

Sharks face a number of threats including shark finning, which has killed over 28 million sharks.

Despite their fearsome reputation, sharks are rarely dangerous to humans. In fact sharks have far more to fear from us than we do from them. There are hundreds of different species of shark in the world, yet only a small number have ever been known to attack humans. Most feed on shellfish and small fishes and avoid humans. In fact, you are more likely to be attacked or killed by a pet dog or even a falling coconut than a shark!

~~~~RESEARCH HIGHLIGHTS~~~~

Ongoing collapse of coral reef shark populations
On coral reefs, sharks are apex predators and are vital to maintaining a healthy reef ecosystem, yet despite conservation management efforts in the form of no-take zones, populations of two of the most abundant reef sharks species on the Great Barrier Reef, Australia are suffering ongoing collapse. No-take zones are difficult to enforce and offer no protection for sharks, even on one of the world's most well-managed ecosystems. Researchers found that there was an order of magnitude fewer sharks on reefs that were fished when compared to no-entry management zones that encompass just 1% of reefs. Whitetip and gray reef sharks showed ongoing decline in abundance of 7% and 17% per year respectively. (Robbins et al. 2006)

Ongoing Collapse of Coral-Reef Shark Populations

Reef sharks, as apex predators, play a major role in the health of reef ecosystems, but are in decline, mainly due to fisheries overexploitation and the conservative life-histories of sharks. Conservation management depends on no-take zones on marine reserves, yet these measures are failing when compared to no-entry zones. The most abundant species, Whitetip (*Triaenodon obesus*) and gray (*Carcharhinus amblyrhynchos*) reef sharks were studied recently on the Great Barrier Reef and the resulting population viability models predict a steep decline in the abundance of 7% per annum for whitetip sharks and 17% per annum for gray reef sharks, strongly suggesting that current management is inadequate for protecting reef sharks, even on the Great Barrier Reef, one of the best managed reef ecosystems in the world.

Reef sharks, as apex predators, play a major role in the health of reef ecosystems, but are in decline, mainly due to fisheries overexploitation and the conservative life-histories of sharks.

On the Great Barrier Reef, there are four levels of coral reef management zones, which can be seen in figure 45:

1. No-entry zones, which are strictly enforced exclusion areas comprising just 1% of the total reef area.
2. No-take zones, which allow fishing boats to be present but cannot legally be fished (there is a moderate level of illegal fishing). These comprise 30% of the total reef area.
3. Limited-fishing zones which have tight restrictions on the permitted fishing gear and comprise 4% of the reef area.
4. Open-fishing zones with fewer restrictions on line fishing and comprise 60% of the reef area.

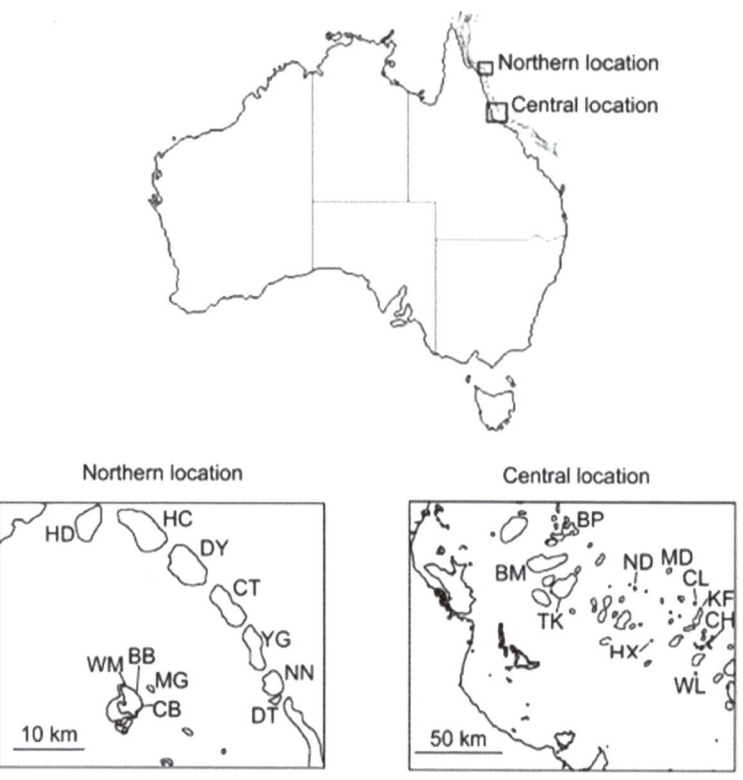

Figure 45. Location of Great Barrier Reef Underwater Visual CensusesNo-entry zone reefs surveyed were Carter (CT) and Hilder (HD) reefs; no-take zone reefs were Barnett Patches (BP), Coil (CL), Detached (DT), MacGillivray (MG), No Name (NN), and Wheeler (WL) reefs; limited-fishing zone reefs were Bommie Bay (BB), Crystal Beach (CB), Myrmidon (MD), Needle (ND), Trunk (TK), and Washing Machine (WM) reefs; open-fishing zone reefs were Britomart (BM), Chicken (CH), Day (DY), Helix (HX), Hicks (HC), Knife (KF), and Yonge (YG) reefs. Current zonation of BB, CB, MG, and WM reefs was implemented in 1983; other listed northern reef zones were implemented in 1992. All listed central reef zones were implemented in 1987. (Taken from Robbins et al. (2006) Ongoing Collapse of Coral-Reef Shark Populations. Current Biology. Vol 16, Issue 23. Pp. 2314-2319).

The abundance of reef sharks in these terms were compared to samples from the southern atoll of the Cocos (Keeling) Islands in the Indian Ocean, which may be one of the last pristine reefs in the world, with no recorded history of commercial shark fishing and negligible recreational shark fishing.

The study found that they were more sharks inside no-entry zones, a figure similar to the shark populations in the Cocos Islands but there was an 80% reduction in whitetip and 97% reduction in gray reef sharks in areas with the fewest fishing restrictions on the Great Barrier Reef (fig 45). Even on no-take reefs, where fishing is illegal but fishing boats are permitted to anchor, shark numbers were heavily depleted and remarkably similar to the legally fished zones. No-entry zones are clearly more effective than no-take zones.

> For shark conservation, no-entry zones are clearly more effective than no-take zones.

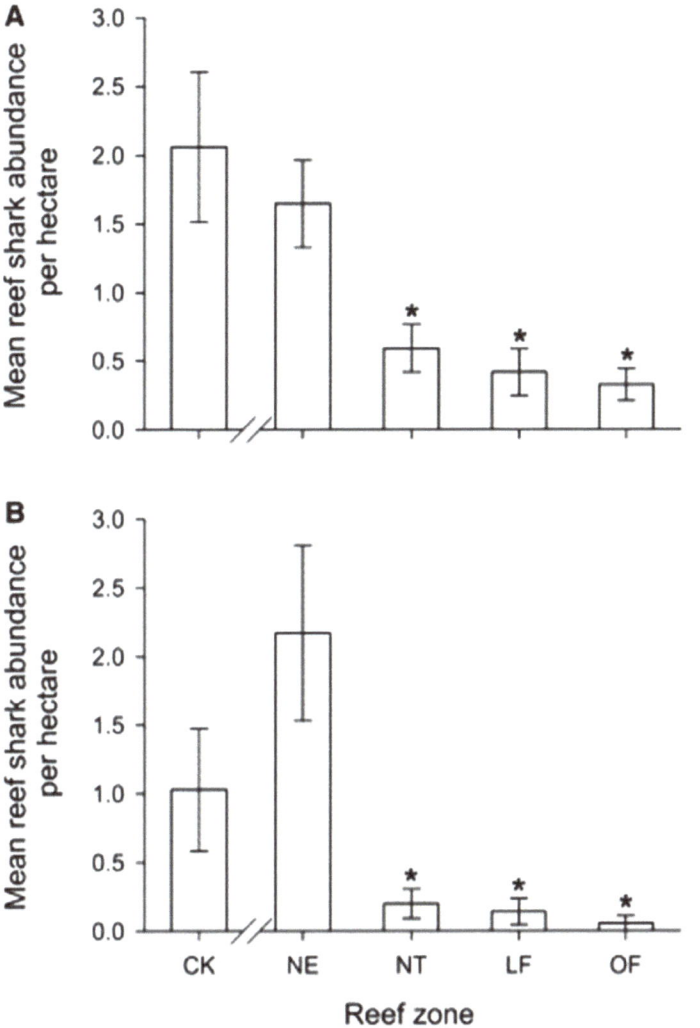

Figure 46. Abundance of Reef Shark on Coral-Reef Fronts. Mean abundance of whitetip reef sharks (A) and gray reef sharks (B) estimated through underwater visual surveys at the Cocos (Keeling) Islands (CK) and at no-entry (NE), no-take (NT), limited-fishing (LF), and open-fishing (OF) management zones on the Great Barrier Reef, Australia. Error bars represent standard errors. Seventeen surveys were undertaken at the Cocos (Keeling) Islands; 19 were undertaken in each of the NE, NT, and LF zones, and 23 were undertaken in OF zones. Asterisks denote management zones that significantly differ from no-entry (NE) zones; $p < 0.005$. Reef shark abundances do not significantly differ among no-take, limited-fishing, and open-fishing zones; $p > 0.7$ for each comparison. Taken from Robbins et al. (2006) Ongoing Collapse of Coral-Reef Shark Populations. Current Biology. Vol 16, Issue 23. Pp. 2314-2319

Illegal fishing in no-take zones is high, despite the policing of inshore reefs.

One possible explanation for this is that illegal fishing in no-take zones is high, despite the policing of inshore reefs. This is a problem in marine reserves throughout the world. Not only could direct overfishing of sharks be causing a decline, but secondarily , the fishing of their prey species, may force sharks to seek prey on the reefs of no-entry zones where neither fishing or entry to the area is permitted.

A population viability analysis of the data showed that its current population trends continue unabated, the abundance of whitetip reef sharks and gray reef sharks on legally fished reefs will be reduced to just 5% and 0.1% respectively within 20 years. Low shark population densities will compound the problem further due to difficulty in finding mates and the effects of low genetic variability.

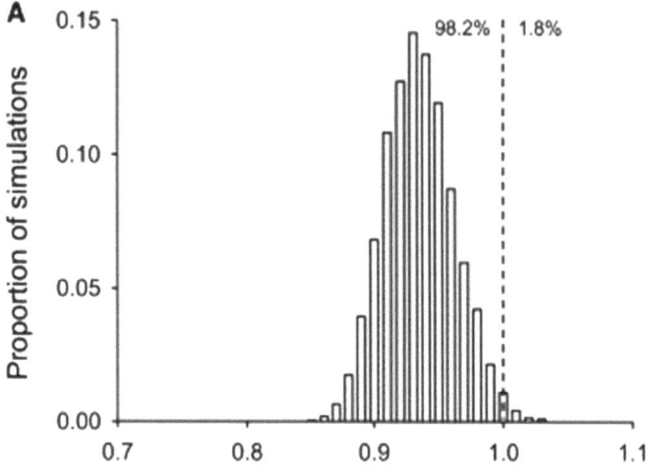

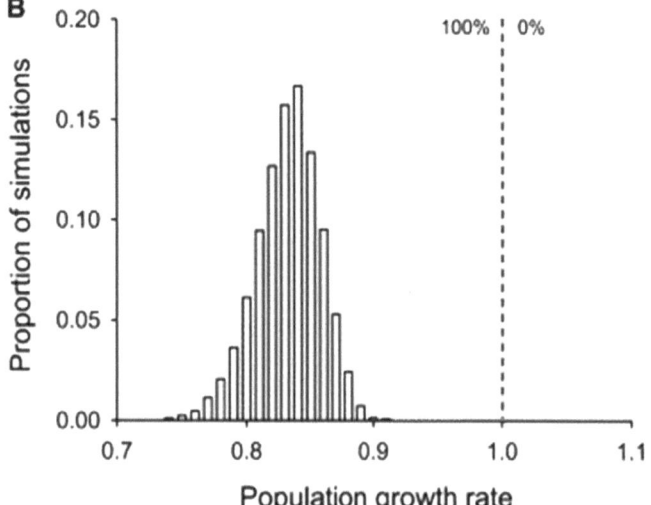

Figure 47. Estimated Population Growth Rates of Reef Sharks. Uncertainty distribution of population growth rates of whitetip reef sharks (A) and gray reef sharks (B) generated from 10,000 replicate Monte Carlo repeated simulations. Percentages indicate the frequency of simulations that project declining (left of dashed line) versus increasing (right of dashed line) populations. (Taken from Robbins et al. (2006) Ongoing Collapse of Coral-Reef Shark Populations. Current Biology. Vol 16, Issue 23. Pp. 2314-2319).

The study shows that the minimum change in mortality necessary to produce a median estimated population growth rate of 1.0 (i.e., population stability) was calculated for each species. Analyses indicate that reductions in annual mortality by one-third (36%) for the whitetip reef shark and one-half (49%) for the gray reef shark would be required to halt these ongoing declines. A review of conservation management strategies in view of this threat to coral reef sharks is suggested, with an increase in no-entry areas, which seems to offer some protection to these sharks.

Turtles

If you remember back to our discussion of turtles in chapter five, you may recall that tortoises and turtles are the oldest of all living reptiles, appearing some 200 Ma. By 140 Ma, they were the largest animals in the ocean. Apart from the leatherback turtle, *Dermochelys coriacea*, almost all are confined to tropical and sub-tropical waters, with the largest abundance found around coastal areas and coral reefs. Marine turtles breed on remote beaches and islands and usually return to their natal beach in large numbers. Their eggs are incubated in nests and the gender and sex ratio of the hatchlings is dependent on the temperature of the nest. Researchers have demonstrated a pivotal incubation temperature of 28.8^0C in the sex ratio of green turtles, *Chelonia mydas*, at Ascension Island in the South Atlantic – one of the most important green turtle rookeries in the Atlantic. Nest temperature, air temperature and sand temperature, as well as the length of incubation determine the sex of the hatchlings, with females favouring higher temperatures than males. Turtles face a number of threats, which we will discuss in detail in chapter eighteen.

Fig 48. Turtles, especially hatchlings face a number of threats.

~~~~RESEARCH HIGHLIGHTS~~~~

Diving behavior of Hawksbill turtles

Time-depth recorders were attached to Hawksbill turtles, *Eretmochelys imbricata*, In Puerto Rico. Dive profiles from four turtles showed strong similarities in diving behavioral diving patterns. During daylight hours, the turtles were active for around 8.4 h per day, surfacing 3.6% of the time. Foraging dives ranged from 8.6 to 14 minutes at a mean depth of 4.7 m and associated with feeding on encrusting sponges. At night, turtles were most inactive, surfacing just 1.8% of the time. From the data on these deep, long night dives, it is suggested that turtles dive aerobically at night, using up stored oxygen. They typically rested on the same ledge for 30-40 minutes, surfaced for air then returned to the same depth, a pattern that was repeated throughout the night. (van Dam and Diez 1996)

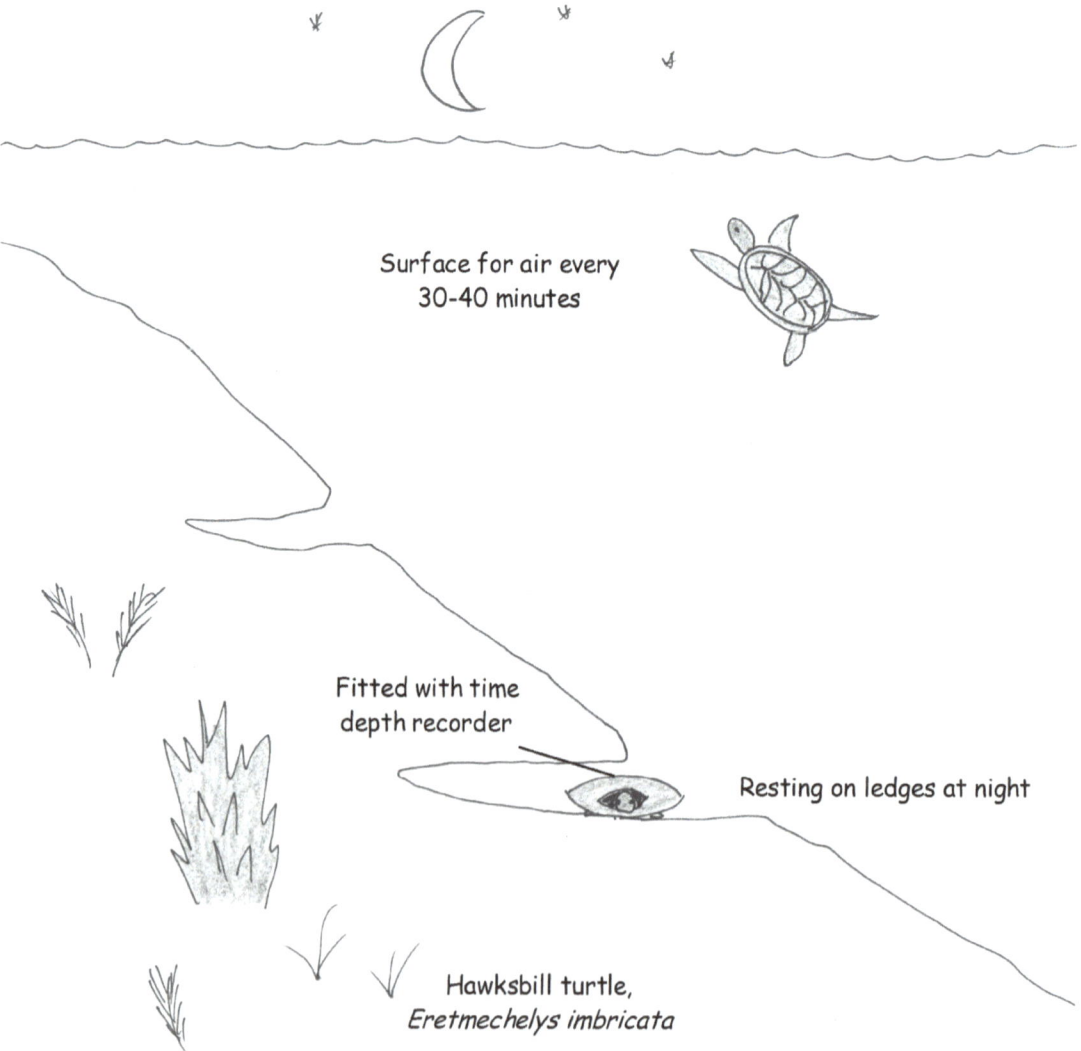

Surface for air every 30-40 minutes

Fitted with time depth recorder

Resting on ledges at night

Hawksbill turtle, *Eretmechelys imbricata*

Fig 49. Diving behaviour of reef turtles. The hawksbill turtle, *Eretmechelys imbricate*, rests at night among crevices on the reef.

Symbiotic relationships

There are many examples of symbiotic relationships between animals on the reef, the most widespread being the relationship between coral polyps and zooxanthellae, as we discussed earlier, and the clownfishes which live in the stinging nematocysts of anemones. Other examples include the cleaning symbiosis that is seen when an animal cleans

Fig 50. A reef cleaning station. There are many symbiotic relationships among animals on the reef, including the cleaner wrasse and its various 'clients'.

ectoparasites (external parasites) from another fish. We came across this earlier when we discussed fish diversity. The first gets a free meal and the other gets rid of his irritating 'guests'. Cleaners include several species of shrimp and small fishes such as butterfly fish and the best known cleaner, the cleaner wrasse, *Labroides* sp. They are generally solitary animals, brightly coloured and equipped with pointed snouts and beaks to enable them to pick off parasites from their hosts.

~~~~RESEARCH HIGHLIGHTS~~~~

Cleaner fish drive diversity on coral reefs
The processes that drive reef diversity are not fully understood, although cleaner fish are thought to be a key species that drive diversity by the removal of large numbers of parasites. The cleaner fish *Labroides demidatus* was found to have major effects on fish activity patterns and may indirectly affect fish demography. Experiments where *L. demidatus* was excluded for 18 months showed half the diversity of fish species and one-fourth the abundance of individuals. It seems that many fish choose reefs based on the presence of cleaner fish. This small fish can strongly affect the movements, habitat choice, activity and local diversity and abundance of a wide variety of reef fish.
(Grutter et al. 2003)

Cleaning usually takes place at designated cleaning stations at various strategic locations around the reef. Cleaners wait at the station and their 'clients' approach to be cleaned of parasites from their gills, skin, fins and even inside their mouth. Cleaners perform a vital service and there is often a queue! It is an important aspect of the ecology of the reef.

There are many symbiotic relationships between animals on the reef.

Fig 51. Cleaner fish. Many species on the reef use the services of cleaner fish, which may be an important driver for reef diversity.

One intriguing and poorly understood feature of this symbiotic relationship is how clients recognize cleaners and decide not to eat them. As figure 51 illustrates, there is a variety of fish species that use the services of cleaner fish, including surgeon fishes and moray eels, and they all allow cleaner fish to remove parasites and even invade their mouth and gills.

The colour and body pattern of cleaner fish is thought to signal cleaning behavior to client fish on the reef, particularly the blue colouration pattern seen clearly in figure 30. In fact, Karen Cheney and her colleagues at the University of Queensland, Australia recently showed that cleaner fish are more likely to display blue and yellow colouration when compared to non-cleaner fish. From the perspective of potential client fish, blue contrasts most against coral reef backgrounds, whereas yellow contrasts best against the blue of the surrounding seawater or against black lateral stripes.

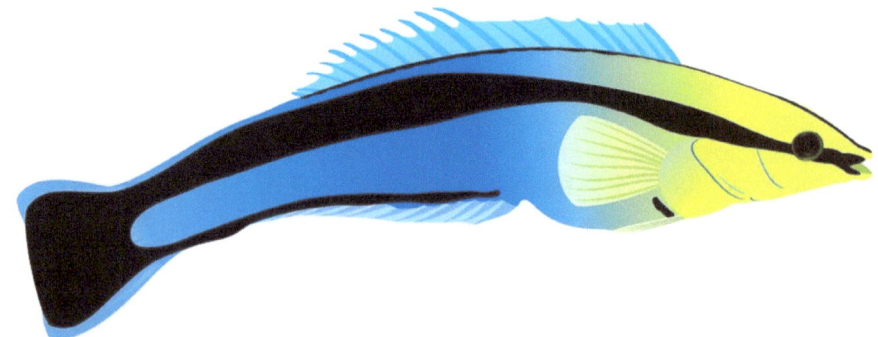

Fig 52. The cleaner wrasse. Cleaner fishes show more blue and yellow colouration than non-cleaner fish.

Remoras are fish whose first dorsal fin has been modified as a sucker. They attach themselves to other fish, often sharks, dolphins and whales. This association enables remoras to feed on parasites and scraps of food from its host. It also acquires the protection of a larger animal. This protective relationship is also seen in the clownfish who hide amongst the tentacles of anemones and in shrimp fishes that hide in the spines of sea urchins.

The threats to coral reefs

Coral reefs have been in existence for some 500 million years, yet these iconic and beautiful ecosystems are threatened because of anthropogenic changes to the atmosphere and ocean chemistry that results in coral death by bleaching and disease, as well as a decline in calcification of coral skeletons caused by an increase in atmospheric CO_2 that leads to ocean acidification. Destruction of coral can occur by physical, chemical or biological factors, either directly by hurricanes, nutrient run-off or predators, for example, or indirectly by such changes as overfishing or climate warming.

> Threats to coral reefs include physical or biological factors, either directly by hurricanes, predators (such as *Acanthaster*) or human disturbance, for example, or indirectly by overfishing or climate warming.

Already approximately 20% of the world's reefs are lost and some 26% are under imminent threat. In many cases, we already have the conservation science and knowledge to address these problems, but because of other priorities in society, economic pressures and the obstructive legal systems of many countries, the practical application of conservation remedies is frustratingly slow.

In a tenth anniversary review of coral reef ecosystems, Janice Lough from the Australian Institute of Marine science in Townsville, Queensland, Australia, reported in the Journal of Environmental Monitoring (2008) that the health of many of the world's coral reefs have already been severely compromised because of human activities. While local-scale impacts can be potentially controlled and even ameliorated

~~~~RESEARCH HIGHLIGHTS~~~~

Extinction risk to coral reefs
The conservation status of 845 zooxanthellae reef-building corals was assessed using the International Union for Conservation of Nature Red List Criteria (IUCN). Of the 704 species assigned conservation status, 32.8% are in categories with an elevated risk of extinction. Declines in abundance are associated with bleaching and diseases driven by rises in sea surface temperatures as well as local-scale anthropogenic disturbances. The Caribbean and western Pacific are areas with the highest risk of extinction for corals. (Carpenter et al. 2008)

with the right conservation management, global human actions (mainly in countries outside the tropics) are threatening the survival of coral reefs.

Moderate warming of the tropical oceans has already resulted in an increase in mass coral bleaching events -- the frequency of which will only increase as global temperatures continue to rise. Ocean chemistry is changing as the oceans absorb part of the excess atmospheric carbon dioxide, and changing weather and ocean circulation patterns will affect coral reefs and their many associated plants and animals. Coral reefs may not disappear entirely, but their appearance, structure and community may well radically change. Lough calls for drastic strategies to control greenhouse gases.

~~~~RESEARCH HIGHLIGHTS~~~~

Extinction and evolution
The mass extinctions in the fossil record were a creative force giving opportunities for evolution and diversification. Today, the impact of human activity is leading to a sixth mass extinction event in the oceans, called the Anthropocene Mass Extinction. This extinction will have unknown ecological and evolutionary consequences and is being caused by the synergistic effects of habitat destruction, overfishing, introduced species, global warming, ocean acidification, pollution and nutrient runoff. Rates of destruction and change are accelerating with sudden phase shifts, rather than a smooth, linear progression. Fundamental changes are required to reverse these trends.
(Jackson et al. 2008)

Many successful examples of coral reef conservation have been described, from the national level to community enforced local action, where protected areas have allowed the regeneration of coral reefs and their associated marine species. Local communities can support coral reef conservation while raising income from tourism -- coral reefs create an annual income in south Florida alone of over US$4 billion (£2.4 billion). With the right motivation, coral reef conservation can be successful in every part of the world.

Global trajectories of decline

The degradation of coral reefs over recent centuries has seen a massive decline in species abundance, diversity and habitat structure due to pollution and overfishing. More recently disease and coral bleaching has caused substantial mortality.

A recent study attempted to predict the global trajectories of the long-term decline coral reefs by reconstructing the ecological histories of 14 coral reef ecosystems worldwide from prehuman times the present (table 2). The status of reefs ranged from pristine to globally extinct. Seven general categories of biota were included and referred to as guilds and paleontological, archaeological, historical, fisheries and ecological data were used in the same analysis (table 2 and figure 53).

> The degradation of coral reefs over recent centuries has seen a massive decline in species abundance, diversity and habitat structure due to pollution, disease and overfishing.

Table 2. Ecological states and criteria used to assess the 14 tropical marine sites analyzed.

Ecological state	Criteria for classification
Pristine	Detailed historical record of marine resource lacks any evidence of human use or damage.
	Example: Fossil coral assemblages
Abundant/common	Human use with no evidence of reduction of marine resource.
	Example: No reduction in size of fish vertebrae in middens or relative abundance of species
Depleted/uncommon	Human use and evidence of reduced abundance (number, size, biomass, etc.).
	Examples: Shift to smaller sized fish; decrease in abundance, size, or proportional representation of species
Rare	Evidence of severe human impact.
	Examples: Truncated geographic ranges; greatly reduced population size; harvesting of pre-reproductive individuals
Ecologically extinct	Rarely observed and further reduction would have no further environmental effect.
	Examples: Observation of individual sighting considered worthy of publication; local extinctions
Globally extinct	No longer in existence.
	Example: Caribbean monk seal

(Taken from John M. Pandolfi et al. (2003) Global Trajectories of the Long-Term Decline of Coral Reef Ecosystems. Science vol. 301.no 5635, pp955-958).

Figure 53 shows how the status of each average ecological guild declined sharply over time, with large animals declining faster than small animals and free-living animals declining faster than seagrasses corals, the architectural builders. During the prehuman period, reefs were defined as pristine (the blue colour in figure 53, A to G).

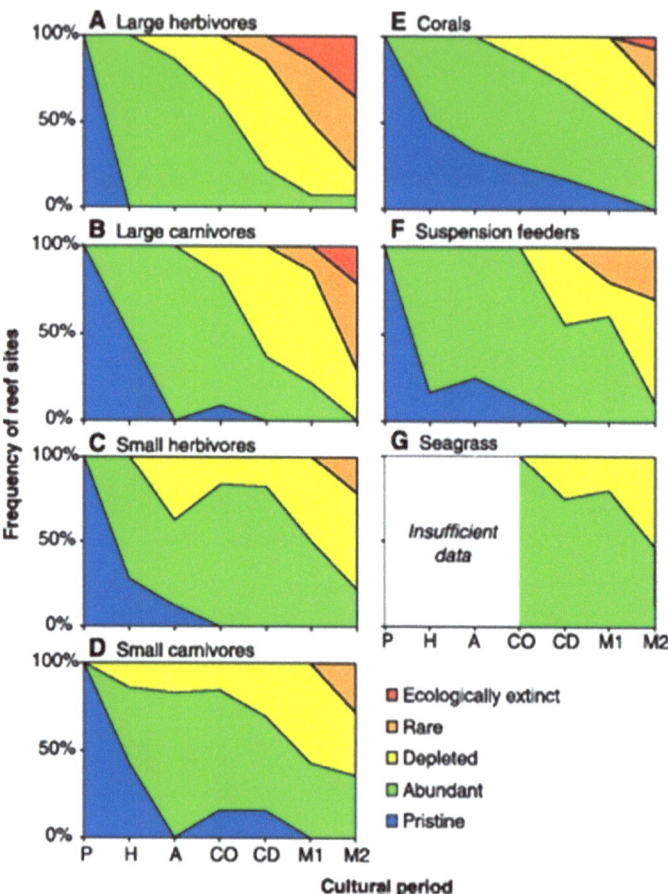

Fig. 53. (A to G) Ecological change in coral reef guilds through time. Time trajectories of ecological condition for each of seven guilds of reef inhabitants (17) expressed as the percentage of regions in each ecological state from 14 regions (16) in the tropical western Atlantic, Red Sea, and northern Australia. Cultural periods (18): P, prehuman; H, hunter-gatherer; A, agricultural; CO, colonial occupation; CD, colonial development; M1, early modern; M2, late modern to present. (Taken from John M. Pandolfi et al. (2003) Global Trajectories of the Long-Term Decline of Coral Reef Ecosystems. Science vol. 301.no 5635, pp955-958).

Fig. 54. PCA of ecosystem degradation based on the ecological state of all seven guilds of reef inhabitants at the 14 reef regions. Only PC1 was significant (17). (**A**) Time trajectories for each reef region over seven cultural periods. Each reef started at a single point to the left in the PCA space that is the pristine ecosystem state (Table 1) (17). Trajectories are mostly monotonic through time, but minor reversals occur in four regions (denoted with an "x" in the filled circle). The hypothetical ecologically extinct state, on the right, is one in which all seven guilds are ecologically extinct. PC1 is interpreted as an axis of historical degradation over time measured in cultural periods. The most important guilds influencing the trajectories of decline are large herbivores and carnivores (20). (**B**) End points

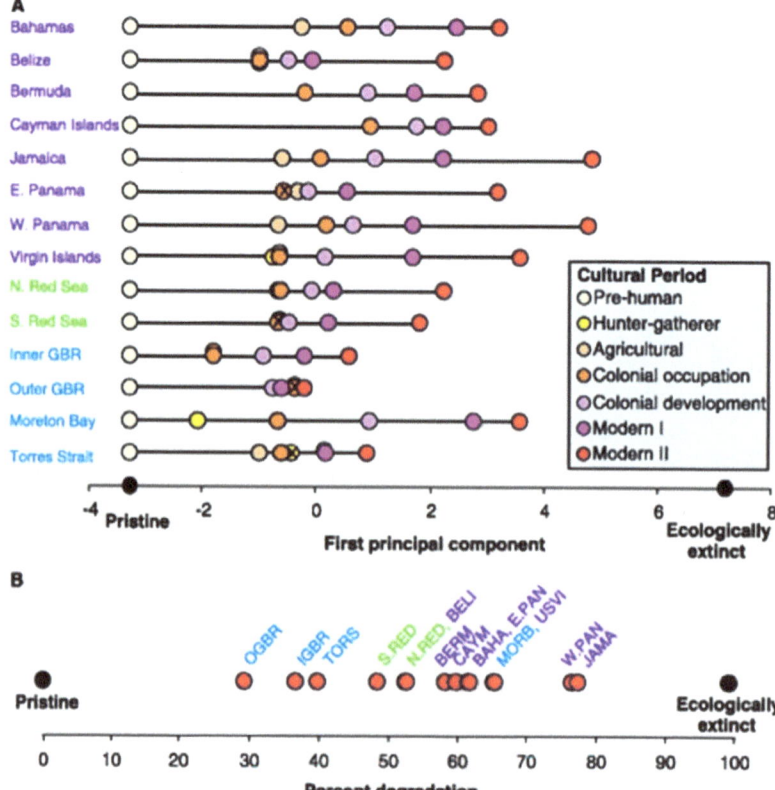

(present ecosystem condition) of the 14 reef regions plotted along an axis of ecosystem degradation measured as the relative distance along PC1 between pristine and ecologically extinct. Oceanic regions are color coded: Australia, blue; Red Sea, green; western Atlantic, purple. OGBR, outer Great Barrier Reef; IGBR, inner Great Barrier Reef; TORS, Torres Strait Islands; S.RED, southern Red Sea; N.RED, northern Red Sea; BELI, Belize; BERM, Bermuda; CAYM, Cayman Islands; BAHA, Bahamas; E.PAN, eastern Panama; MORB, Moreton Bay; USVI, U.S. Virgin Islands; W.PAN, western Panama; JAMA, Jamaica. (Taken from John M. Pandolfi et al. (2003) Global Trajectories of the Long-Term Decline of Coral Reef Ecosystems. Science vol. 301.no 5635, pp955-958).

The Great Barrier Reef is one of the best protected reefs in the world and is close to pristine. However, they are also one-quarter to one-third of the way towards ecological extinction with the reefs of Moreton Bay at the southern end of the Great Barrier Reef is close to extinction as the severely degraded reefs of eastern Panama and the Virgin Islands. Reefs in the western Atlantic have declined more than in Australia or the Red Sea. The overall historical trajectory of reef degradation for all periods is a linear decline, with most reef ecosystems substantially degraded before 1900, likely due to overfishing and pollution -- a fact that has been overlooked with the recent widespread episodes of coral bleaching and disease.

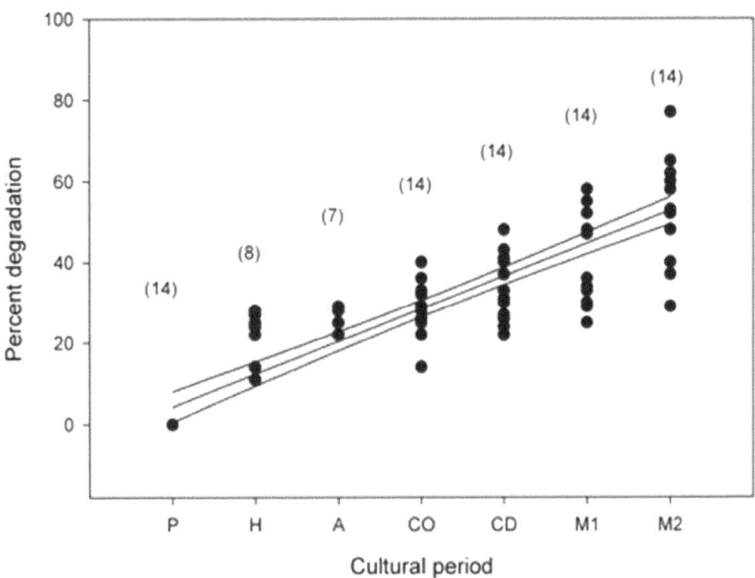

Fig. 55. Percent degradation of 14 reef regions over time. Data for each cultural period are derived from the PCA analysis plotted in Fig. xA as measured along PC1 as the axis of reef degradation. Each point represents percent degradation of a particular site at a particular time. Numbers in parentheses are the numbers of reef regions recorded for each cultural period (*17*). Linear regression is plotted along with the 95% confidence interval. Abbreviations for cultural periods are as in Fig. x. Taken from John M. Pandolfi et al. (2003) Global Trajectories of the Long-Term Decline of Coral Reef Ecosystems. Science vol. 301.no 5635, pp955-958

These historical trajectories may be valuable in predicting future ecosystem decline and implementing the correct conservation management strategies.

Storm damage

Tropical storms and hurricanes break off large chunks of coral, throwing up boulders onto the reef flats. Damage may be substantial but restricted in area within the tropical storm belt except in severe ENSO events. Hurricane damage was well documented by researchers when hurricane Allen hit the North shore of Jamaica, causing damage to reefs at Discovery Bay, particularly to *Acropora cervicornis* and *Acropora palmata*. Damage from increased freshwater due to rainfall during storms can also be significant. Sediment can also overwhelm coral's defenses when storms or human activities create sediment by local disturbance or terrestrial runoff from forest clearance, road building and the like. However, disturbance to coral reefs can open up opportunities for larval settlement and diversity.

> ~~~~RESEARCH HIGHLIGHTS~~~~
>
> **5,000 years of super-cyclones along the GBR**
> High-frequency super-cyclones have raged over the Great Barrier Reef (GBR) over the past 5,000 years, as seen from ridges of detrital coral and shell deposited above the highest tide. They are also found on terraces that have been eroded into coarse-grained alluvial fan deposits. These features are found along 1,500 km of GBR are and the Gulf of Carpentaria, Australia, and were formed by extremely intensive storms with recurrence intervals of two to three centuries -- an order of magnitude higher than has been estimated previously (which was once every several millennia). These storms could be instrumental in shaping the character of coral reef and rainforest communities. (Nott and Hayne 2001)

Hurricanes and coral bleaching has been linked to changes in coral recruitment in Tobago, in the Caribbean. In the years following hurricanes, tropical storms and bleaching events, coral recruitment was reduced when compared to normal years. Following hurricane Ivan in 2004 and the 2005-2006 bleaching event, coral recruitment was limited to only 2% of colonies studied. Despite the huge disturbance from hurricanes, corals are still recruiting, albeit in low numbers.

Hurricanes may also benefit bleached corals because the passage of a hurricane can alleviate thermal stress on coral reefs by cooling the surface temperatures. Hurricane-induced cooling was found to be related to differences in the extent and recovery time of coral bleaching between the Florida Reef Tract and the US Virgin Islands during the Caribbean-wide 2005 bleaching event.

> Disturbance to coral reefs can open up opportunities for larval settlement and diversity.

Fig 56. Tropical storms. Storms can cause catastrophic damage to coral reefs, but can also have beneficial effects.

Bleaching events

Related to global warming is the increase in bleaching events where the zooxanthellae leave their host due to high levels of UV radiation. The coral heads are bleached white when they do not contain zooxanthellae although the sides of the coral may survive. Bleaching occurs when temperatures rise acutely by $3\text{-}4^0$C or during a prolonged rise of $1\text{-}2^0$C. This has been seen in corals monitored in areas such as the Maldives, the Caribbean, Bermuda and Indonesia.

> Bleaching events occur when zooxanthellae leave their host and can occur when temperatures rise acutely by $3\text{-}4^0$C or during a prolonged rise of $1\text{-}2^0$C.

Sometimes individual corals can recover from bleaching events, but are sadly often killed and the reef as a whole may take several years to recover. The severity and frequency of bleaching events has increased since the 1960's in many important species of reef building coral such as *Acropora palmata, Posillopora, Monastrea, Millepora* and *Helipora*.

> The exact mechanisms of coral bleaching are not fully understood but are thought to result from physiological damage to the host cells and/or symbionts, triggered by various environmental factors.

~~~~RESEARCH HIGHLIGHTS~~~~

UV radiation and coral bleaching

Researchers Gleason and Wellington experimented with the reef building coral *Agaricia agaricites* where they transplanted separate colonies at 3 m and 24 m depths. The colonies were subjected to three light regimes for 72 hours. At 3 m depth, larvae showed lower survivorship when exposed to the full light regime than larvae that were partially shielded with acrylic screens. They found that by monitoring zooxanthellae abundance that corals maintained UV screening substances proportional to depth. Deeper corals were unable to cope with higher UV light levels and consequently suffered bleaching. So not only increases in temperature but increases in UV radiation can have catastrophic effects on corals. Continued degradation of the ozone layer means that UV radiation may cause problems for corals as well as humans in the coming years.
(Gleason and Wellington 1993)

In 1987, there was a worldwide bleaching event that was particularly severe in the Caribbean area. Several research studies suggest that this could be a cyclical phenomenon that is increasing in more recent years. Scleractinians are especially affected, for example, *Acropora, Pocillopora, Monastrea* and *Helipora*. Disease following bleaching events make corals more vulnerable and include two main diseases prevalent in the Caribbean; black band disease and white band disease caused by bacteria. We will look at these further in the next section.

The exact mechanisms of coral bleaching are not fully understood but the breakdown of the symbiotic relationship between the coral polyp and it is zooxanthellae is thought to result from physiological damage to the host cells and/or symbionts, triggered by various environmental factors. Cellular oxidation appears to play an important role and oxidative stress occurs during an increase in the radiance and temperature -- the two main factors involved in bleaching. Experiments were conducted using the reef building coral *Pocillopora verrucosa* transplanted between 5 m and 20 m in depth at Grande Glorieuse Island in the Indian Ocean (this area suffers little from direct human-induced change). The study demonstrated that colonies transplanted upward toward the sea surface showed cellular damage and an increase in oxidative stress, whereas the downward transplanted colonies showed no associated alterations.

~~~~RESEARCH HIGHLIGHTS~~~~

Corals eat zooxanthellae when stressed!

Another cellular mechanism being investigated by marine biologists in an effort to understand coral bleaching is symbiophagy, a mechanism whereby the symbiotic zooxanthellae is literally consumed by the host coral polyp. Results of recent experiments show that during stress, the vacuole membrane normally used for nutrient exchange in the coral polyp transforms itself into an organ of digestion and proceeds to consume the zooxanthellae. The death of the zooxanthellae results in bleaching and this may be an ancient mechanism which evolved from a more general primordial mechanism that destroys foreign microbial invaders of the cell, termed xenophagy.
(Downs et al. 2009).

Biological disturbance

Destruction of coral includes the action of predators such as crabs, molluscs, fish and echinoderms, particularly the crown of thorns starfish, *Acanthaster planci.*

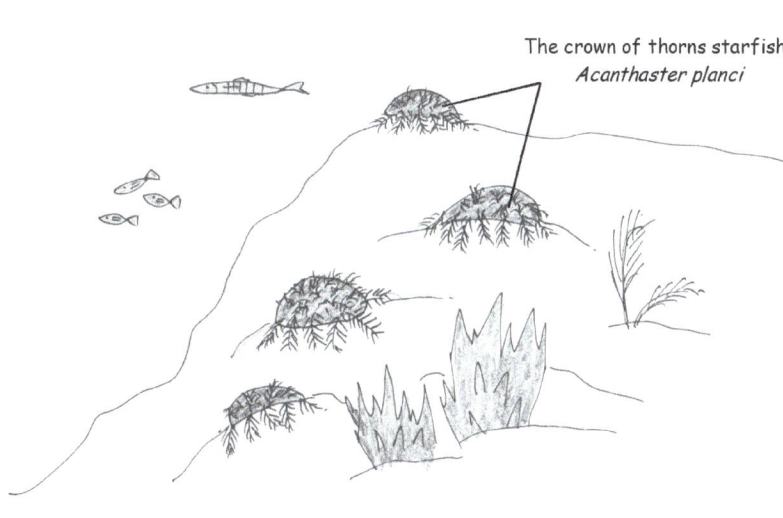

The crown of thorns starfish
Acanthaster planci

Fig 57. The crown of thorns starfish, *Acanthaster planci*, destroying the reef.

Boring organisms such as cyanobacteria and other algae can cause substantial damage to reefs and some fish are specialists at biting coral, for example, the box fish, *Ostracion*, the trigger fish, *Balistes* and file fish, *Monacanthus.*

~~~~RESEARCH HIGHLIGHTS~~~~

***Acanthaster planci* causes coral death in Papua New Guinea**
Acanthaster outbreaks are still a significant threat to coral reefs, despite the attention given to other threats such as global warming. Dramatic increases in abundance of *Acanthaster* were seen in September 2005 on reefs in Bootless Bay, Central Province, Papua New Guinea, where densities of the starfish reached 162 individuals per ha. They caused extensive coral mortality. This outbreak alone killed more than 55% of live corals, reducing coral cover from 42.4% to just 19.1% by March 2006. This damage altered the composition of the coral as well as the physical structure of the reefs habitats affected.
(Pratchett et al. 2009)

Acanthaster, is a specialist coral eater, which can kill more than 90% of the corals on the reef. This was seen in 1957 in Guam then later on the Great Barrier Reef and it subsequently spread outwards. The first plague in the sixties destroyed many reefs in Micronesia and the Great Barrier Reef with later outbreaks in the 1980's that killed 25% of old *Porites* corals. Some reefs showed little recovery after several years.

TROPICAL MARINE ENVIRONMENTS

Black band disease is also a significant threat to coral reefs. Black band disease is a collection of pathogenic microorganisms that predominantly affect the reef building scleractinian corals on reefs around the world. It was discovered first in Belize and was thought to be caused by blue-green algae, *Oscillatoria membrancea* (Antonius 1973) but now a consortium of hundreds of microorganisms is thought to infect the coral and there is still considerable debate as to the composition of microbes that make up the disease. Black band disease infects coral by destroying tissue and can spread at alarming rates of 3 mm to 1

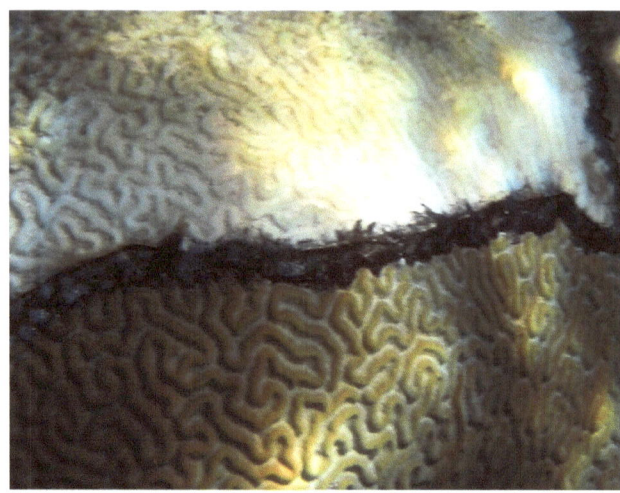

Fig 58. Black band disease. Seen here on a Caribbean brain coral.

cm per day. Bacteria appear as a dark red or black mat, depending on vertical position of the microorganisms within the mat. It is seen between healthy tissue and the exposed skeleton of the coral.

Both photosynthetic and non-photosynthetic bacteria appear to act synergistically. *Beggiatoa* bacteria, the sulpher-oxidsing bacteria that we met colonizing hydrothermal vents in chapter three are also present in the black band disease consortium. In one study, the disease was seen most abundantly in a sewage impacted reef site in St. Croix, US Virgin Islands. Black band disease is thought to be composed of a consortium of bacteria, which vary biogeographically in association with environmental differences, and thrive where there is degraded water quality -- especially due to anthropogenic impacts.

White band disease affects two species of scleractinian coral in the Caribbean, *Acropora cervicornis* and *A. palmata*. There are two types of white band disease with different patterns of tissue degradation, although it is poorly studied. Other coral diseases include brown band disease, yellow band disease and white plague.

> Diseases that affect corals include black band and white band disease, as well as brown and yellow band diseases and white plague.

~~~~RESEARCH HIGHLIGHTS~~~~

Resistance to infections in soft coral
In the Gulf of Eilat, northern Red Sea, hard and soft corals were examined for antimicrobial activity. While the majority (83%) of the Red Sea soft corals showed appreciable antimicrobial activity against marine bacteria, the hard corals did not. This suggests that soft and hard corals have developed different means to combat potential microbial infections, with soft corals using chemical defence through the production of antibiotic compounds. Hard corals use some other mechanism.
(Kelman et al. 2006)

Resistance to infections in staghorn corals
The abilities of two critically endangered (IUCN Red List) species of staghorn coral in the Caribbean, *Acropora cervicornis* and *A. palmata* were investigated for natural white band disease resistance. Results show that 6% of staghorn corals are resistant to white band disease and this is the first evidence of disease resistance in scleractinian corals. It explains why pockets of *Acropora* have been able to survive epidemics and may be important in conservation management efforts.
(Vollmer et al. 2008)

68

Increasing sea surface temperatures also affect the dynamics of disease at large spatial scales. John Bruno and his team at the University of North Carolina used a new high resolution satellite dataset of ocean temperatures and compared it with data from annual surveys of 48 coral reefs that gave information on coral cover and coral diseases. They found a highly significant relationship between the frequencies of warm temperature anomalies and of white syndrome, an emergent disease of Pacific reef building corals. On high cover reefs, white syndrome outbreaks followed warm years. This correlation between warm sea temperatures and coral disease is worrying considering the projected trend in global warming.

> The correlation between warm sea temperatures and coral disease is worrying considering the projected trend in global warming.

Natural disease resistance in threatened staghorn corals

In the Caribbean, diseases have caused widespread destruction of key species such as the herbivorous sea urchin, *Diadema antillarum* and two important shallow water, reef building corals, the staghorn coral, *Acropora cervicornis* and the elkhorn coral, *A. Palmate*. Corals have died off in huge numbers due to White Band Disease (WBD), which only affects Caribbean *Acropora*. They are now listed as critically endangered on the IUCN Red List and this decline of coral reefs in the Caribbean has resulted in a dramatic shift from coral to macroalgal dominance.

However, some researchers have discovered a potential for natural resistance to WBD in the staghorn coral. In fact, some 6% of staghorn coral genotypes are resistant to WBD and may explain why pockets of *Acropora* have been able to survive WBD epidemics.

White Band Disease is seen as a rapidly advancing band of white diseased tissue, as shown in figure 59a). It is transmitted via direct contact and other vectors such as the corallivorous snail *Corallophyllia abbreviate*. Histological and genetic data suggest that the pathogen that causes WBD in the Caribbean is bacterial and the bacterium *Vibrio charcharia* appears to be associated with WBD type II, seen in the Bahamas.

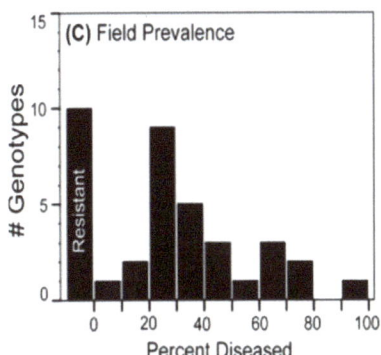

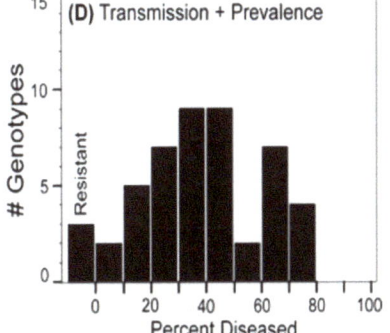

Fig 59. Resistance to White Band Disease (WBD) in the staghorn coral *Acropora cervicornis*.
a)The appearance of WBD
b)The transmission of the disease in experiments
c)The prevalence of the disease in nature
d)The transmission and prevalence of the disease
(Taken from Steven V. Vollmer and David I. Kline (2008) Natural Disease Resistance in Threatened Staghorn Corals. *PLoS ONE* 3(11)).

A study in Panama tagged a total of 106 staghorn coral colonies in four reefs (i.e. populations) and determined their unique genotypes, with 46% of the sample showing unique staghorn coral genotypes (shown in table 3). Each population displayed a relatively high number of genotypes

Table 3. Number of staghorn coral samples (N), unique genotypes (G), and the ratio of genotypes per sample (G/N) per population along with clonal diversity (D_s), its evenness (E), and the number of genets with clones (N_{cg}). (Taken from Steven V. Vollmer and David I. Kline (2008) Natural Disease Resistance in Threatened Staghorn Corals. *PLoS ONE* 3(11)).

Population	N	G	G/N	D_s	E	N_{cg}
Punta Caracol	24	7	0.292	0.558	0,307	3
Casa Blanca	35	12	0.343	0.699	0.260	4
Crawl Cay	23	13	0.565	0.850	0.411	3
Salt Creek	24	17	0.708	0.938	0.584	3
Total	106	49	0.462			

Experiments were conducted in parallel across the four sites in order to study WBD resistance in the different staghorn coral genotypes. Corals were grafted with fragments of WBD (fig 59b) and the successful transmission of the disease averaged 45.5%. Only 3.3% of the control fragments developed WBD, due to poor transmission of the infection.

The 49 different staghorn coral genotypes showed a wide variety of susceptibility to the disease, with most of them getting the disease to some degree, a few were highly susceptible to it (fig 59b). Important and statistically significant findings from these experiments showed that five genotypes did not contract WBD, despite repeated attempts to infect them. The five resistant genotypes came from two of the four reefs studied, Salt Creek and Crawl Cay in Panama.

Around 28.2% of the tagged genotypes were infected with WBD in the field studies (fig 59c), lower than the 45.5% in the *in situ* transmission experiments (fig 59b). The combined data from the in situ transmission experiments and field surveys showed a conservative index of the susceptibility of these corals to the disease (fig 59d). The genotypes that were resistant in the field will also resistant in the transmission experiments, with an estimate of 6% of genotypes being naturally resistant to WBD.

These experiments show that over evolutionary time, some corals have become resistant to disease and this is important for their survival. However, Staghorn corals rely on local asexual fragmentation and if this disease resistance is to spread, the resistant genotypes will have to successfully disperse and recruit their coral larvae. Worryingly, Staghorn corals, the Caribbean *Acropora* populations, appear to be experiencing recruitment failure. The gene flow in other corals such as *A. cervicornis* and *A. palmata* seems to be restricted to spatial scales less than 500 km

> Over evolutionary time, some corals have become resistant to disease and this is important for their survival.

or less, perhaps just 2-5 km, which is not sufficient to allow diseased populations to recover via natural recruitment following outbreaks of WBD. The conservation of these reefs may require the propagation and transplantation of WBD resistant genotypes into damaged reefs.

Human disturbance

Human activities which are destructive to coral include dredging and mining, causing destruction by siltation of the reefs such as in Bermuda and by direct mining of coral as in Male, in the Maldives. Land clearance causes soil erosion and can affect reefs up to 20 km offshore. Pollution by sewage that occurred in Hawaii has caused eutrophication (excessive nutrition), red tides by phytoplankton and the death of coral caused when the human population increased, particularly in the 1960's and 1970's.

> **~~~~RESEARCH HIGHLIGHTS~~~~**
>
> **New perspectives on aquarium fish trade**
> The French Polynesian Islands have the potential to develop a marine aquarium fish business by capturing coral reef fish at the larval stage using crest nets. The larvae are then reared in a query before being sold on the world market for aquarium fish. This avoids environmental damage, especially in south-east Asia where, it is suspected, that fish catches are still executed with cyanide-based toxic products. The use of crest nets also limits the stress on fish larvae and increases survival rates above those of fish captured at the adult stage.
> (Lecchini et al. 2006)

Coral was replaced by algae until corals began to recover following the relocation of the sewage outfall further out to sea. Atolls such as Tarawa in the Gilbert Islands are also susceptible to sewage pollution in lagoons. Tourism may cause local damage by shell collecting, diving and reef walking as well as wider damage by clearance of mangroves for development and the siltation caused by building hotels and roads such as the Cape tribulation road in Queensland.

Marine debris such as derelict fishing gear and litter accumulate and cause damage to shallow coral reefs, particularly in the northwestern Hawaiian Islands, due to the characteristics of ocean circulation patterns that we met in chapter two. One recent study examined previously cleaned back reef and lagoonal reefs at two locations in the northwestern Hawaiian Islands and found that marine debris accumulated more in lagoonal reef areas. Using satellite images, the annual debris accumulation on the reef was estimated to be 52 metric tons.

Overfishing on coral reefs is another principal threat to reef structure, resilience and species diversity. Coral reef fisheries are generally held to be unsustainable, yet a recent study showed the astonishing scale of exploitation. Over half (55%) of the 49 island countries examined were overexploiting their coral reef fisheries, with total landings of coral reef fish currently 64% higher than can be sustained. This has enormous impacts on coral reef ecosystems.

The environmental impacts of artisanal fishing gear on coral reef ecosystems can be considerable. In southern Kenya, researchers evaluated which types of gear had the greatest impact on coral reef

> **~~~~RESEARCH HIGHLIGHTS~~~~**
>
> **Declines in Caribbean reef fish abundance**
> Profound ecological changes are occurring on coral reefs throughout the tropics, but the reduction in coral cover and an increase in algae, particularly in the Caribbean. While the decline in the abundance of large Caribbean Reef fishes is probably due to overexploitation over hundreds of years, the effect of the degradation of reef habitats on reef fish assemblages is still being investigated. The study analysed data obtained from 218 reefs across the Caribbean between 1955-2007, and found that although reef fish have been declining steadily in all sub-regions of the Caribbean basin at rates of 2.7% to 6% loss per year, the recent decline in fish abundance across all trophic groups indicate that Caribbean fishes have begun to respond negatively to recent habitat degradation.
> (Paddack et al. 2009)

biodiversity. The gear included in the study were large and small traps, gillnets, beach seines, hand lines and spear guns. Beach seines, spears and gillnets were found to cause the most direct physical damage to corals. Beach seines landed the highest percentage of juvenile fish (68.4%), which was significantly higher than any other gear and had a considerable impact on reef biodiversity by growth overfishing.

Cyanide fishing, which is illegal in most countries, has been used commonly in the Philippines to stun and collect tropical marine fish aquarium and food trades since the early 1960s. It is damaging coral reefs irreversibly. Even brief exposure to very small amounts causes mortality susceptibility to disease in corals and their zooxanthellae as well as anemones and other reef species.

> Human activities are causing widespread damage to corals reefs around the world.

Conservation efforts are aimed at detecting cyanide (NaCN) in the marine environment. Many of these techniques are time-consuming and relatively insensitive, although new techniques are being developed all the time.

Another troublesome method of fishing is dynamite blasting on coral reefs. This highly destructive practice is seen throughout south-east Asia and creates persistent rubble fields with low coral cover and reduce fish abundance. Explosives are used to catch reef fish – known as fish bombing. How frequently this practice is used is largely unknown, but efforts are being made to develop a detection system capable of distinguishing underwater explosions from background noise, and then by triangulation, locate the source of the blast so that the perpetrators can be caught.

~~~~RESEARCH HIGHLIGHTS~~~~

Coral reveals traces of past agriculture
Porites coral cores from Bali, Indonesia, exposed to high levels of fertilisers in agricultural run-off were collected 30 km offshore. Organic nitrogen preserved in the skeleton reveal the history of nitrogen enrichment in coastal waters over the past 30 years and suggests that the intensification of Western-style agricultural practices since 1970 have been contributing to the degradation of coastal coral reefs.
(Marion et al. 2005)

Efforts are being made to find ways to stabilize these rubble fields and stimulate natural recovery in fish and coral populations with some success. In one experiment in a marine protected area in the Philippines, a 20-year-old rubble field was stabilized using plastic mesh. Within three years fish abundance was similar to an adjacent healthy reef and coral recruitment showed a 63.5% survivorship, compared with 6% on loose rubble.

Urbanisation around Maunalua Bay, Oahu, Hawaii, is well-known for impacting the coral reef ecosystem and affecting important marine organisms through terrestrial run-off, eutrophication and pollution. The Great Barrier

Fig 60. Human impacts. Even sunscreen in low concentrations can cause coral bleaching.

Reef is the world's largest coral reef ecosystem and the best managed for conservation in the world, yet it continues to be degraded from land-based pollution. A recent report illustrates that there are still hot spots for run-off, such as the Fitzroy River Basin, the largest Great Barrier Reef catchment. The study shows that the maximum pollutant concentrations are closely related to the percentage area of croplands receiving heavy rain, although grazing lands contribute majority of the long-term average annual load of most common pollutants.

~~~~RESEARCH HIGHLIGHTS~~~~

Sunscreen causes coral bleaching
Recently, it has been demonstrated that products such as sunscreens have an impact on aquatic organisms, similar to other contaminants. This study conducted in-situ and laboratory experiments to evaluate the potential impact of sunscreen ingredients on hard corals and their symbiotic algae. Results showed that sunscreens cause rapid and complete bleaching of hard corals, even at extremely local concentrations, due to organic UV filters which are able to induce viral infections. This has important implications for areas used for human recreation.
(Danovaro et al. 2008)

Tourism is also having a detrimental effect on coral reef ecosystems throughout the world. Recreational SCUBA diving particularly threatens coral reefs. The reefs at Dahab, South Sinai, Egypt, are among the world's most heavily dived sites, with more than 30,000 dives per year. One recent study compared frequently dive sites to sites with no or little diving. The benthic communities and condition of the corals were examined at the reef crest zone and reef slope zone, and the abundance of coralliverous (fish that eat coral) and herbivorous fish was estimated. Researchers found that the areas subject to intensive SCUBA diving showed a significantly higher number of damaging coral, significantly lower coral cover and higher sedimentation than on the sites that were not dived heavily. Corals at the reef crest were more affected than those of the reef slope, and 95% of the broken colonies were those with branching corals. Fish abundance did not appear to be affected.

Other anthropogenic impact on coral reefs include heavy-metal accumulations from construction, discarded vehicle tyres, paint remains, iron pipe rusts, hydrocarbons, plastic bags, metal and wood remains, bilge water from boats and so on. One study in the Red Sea found high levels of toxic metals; Zn, Cu, Pb, Ni and Cd in fine sediments.

Fig 61. Anthropogenic impacts. Human disturbance is causing widespread damage to coral around the world.

Climate change and coral reefs
Climate change is an important factor affecting coral reefs. The predicted global warming with the accompanying sea level rises will have far reaching impacts for marine life everywhere, but particularly for tropical marine ecosystems. Throughout their evolution, corals have endured a range of temperatures and survived several mass extinction events, but the speed of change due to anthropogenic effects will have unknown consequences for corals and their associated species.

Ocean acidification is a key threat to coral reefs. It reduces the calcification rate of reef building, hard corals and affects the relationship between corals and their zooxanthellae, although there is still much to learn about this mechanism – the consequences of ocean acidification remain unclear. Recent studies indicate that, even at CO_2 stabilization levels as low as 450 ppm, certain sensitive reef building species, such as crustose coralline algae may be pushed beyond their thresholds of growth and survival within the next few decades. Other corals will demonstrate delayed, mixed responses.

~~~~RESEARCH HIGHLIGHTS~~~~

Past decline of coral reefs
In the past couple of decades, the frequency of warming events has intensified alongside widespread coral bleaching, yet it is difficult to predict how coral reefs will react to prolonged environmental pressures. Paleocene and early Eocene shallow water carbonate platforms and palaeoclimate may provide an insight. Between 59 and 55 Ma, three discrete stages in platform development have been identified around the Tethys Sea. Firstly, carbonate platforms of coralgal reefs; secondly a transitional stage where coralgal reefs thrived and in middle latitudes, giving way to larger foraminifera as dominant carbonate producers in low latitudes; finally, newly developed lineages completely took over as carbonate producing organisms in low to mid-latitudes. Rising temperatures led to a stepwise demise of Paleocene coral reefs, giving way to the expansion of larger foraminifera, which dominated the Tethys platforms during the early Eocene.
(Scheibner et al. 2008)

Ocean acidification is a key threat to coral reefs.

At this stage, we simply don't know the extent to which coral reefs will be affected. If nothing changes, anthropogenic induced atmospheric CO_2 concentrations could almost double between 2006 and 2100, leading to increased amounts of dissolved inorganic carbon and CO_2 in surface seawater, resulting in ocean acidification and lower carbonate saturation states. This means that marine calcifying organisms, such as corals, coralline algae, molluscs and foraminifera will have problems producing skeletons and shells at their current rate and significantly change benthic community structure in coral reef ecosystems.

In fact, a recent study shows that there are already surprisingly large decreases in pH, showing acidification, across important carbonate producing regions, such as the Great Barrier Reef.

~~~~RESEARCH HIGHLIGHTS~~~~

Coral reefs may dissolve when CO_2 doubles
Calcification rates in stony corals are expected to decline significantly in the near future due to ocean acidification caused by increasing CO_2 levels and rising sea surface temperatures. In this study calcification rates were calculated from more than 9,000 reef locations, and the resulting maps show that by the time atmospheric partial pressure of CO_2 reaches 560 ppm, all colonies will cease to grow and start to dissolve.
(Silverman et al. 2009)

Ocean acidification will also affect marine fish due to the reduced ability of larvae to detect olfactory cues from adult habitats. As we saw earlier, larvae find suitable reef habitats by smell, and often return to their home reef. One study showed that larval clownfish reared in control seawater with a pH 8.15 discriminated between a range of cues which could help locate reef habitat and suitable settlement sites, but this ability was disrupted when larvae were reared in conditions simulating CO_2 induced ocean acidification of pH 7.8, predicted for the year 2100. They no longer responded to olfactory cues at pH 7.6, also predicted for later in the next century if CO_2 emissions continue unabated. This will have disastrous implications for population sustainability and species diversity.

Mangrove forests

Mangroves are trees or bushes which grow between the intertidal in marine or esturine conditions, covering 60-70% of sheltered tropical coastlines on a variety of substrates; sand, silt, mud, peat and even coral. They cover an area of around 160,000 km^2 and spread at a fast rate over mud flats at around 100 m/yr. They can occur on small islands and atolls but are bounded by latitudes of 35^{0}N and 37^{0}S, replacing salt marshes as coral replaces kelp on rocky substrates. The most extensive mangrove forests or mangals, which refer to the whole mangrove community, occur in the Indo Malayan region. Researchers have suggested that mangroves evolved around the Tethys Sea during the late Cretaceous with species diversity following continental drift, rather than there being a centre of diversity in the Indo West Pacific, as there is for corals. A cycle of expansion and decline of mangals has been described, which is linked to marine regressions and transgressions during the late Quaternary.

> Mangroves cover 60-70% of sheltered tropical coastlines. They cover ~160,000 km^2 and are bounded by latitudes of 35^{0}N and 37^{0}S. The most extensive mangrove forests or mangals occur in the Indo Malayan region.

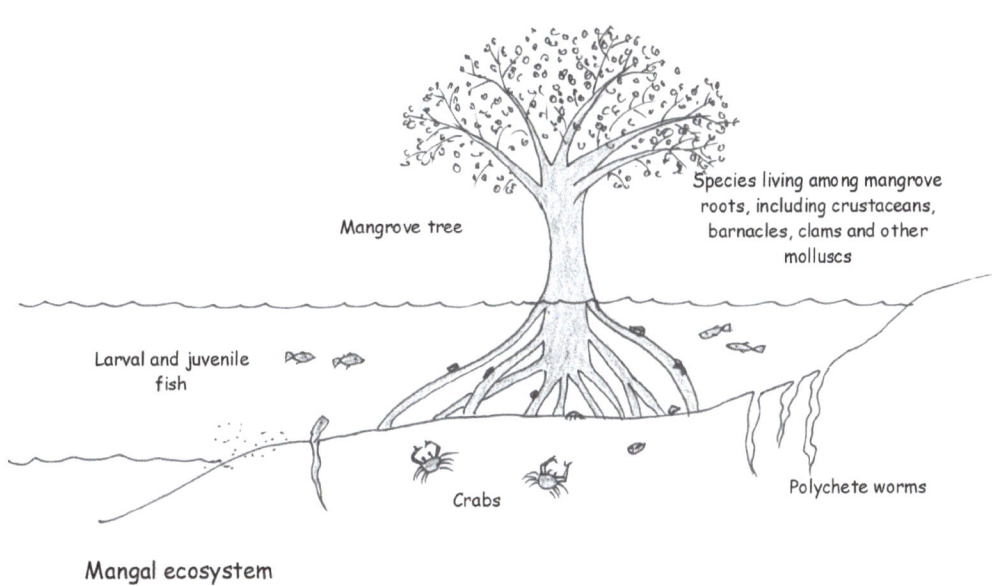

Mangal ecosystem

Fig 62. A mangal ecosystem. Mangals are an important habitat for many animals and a nursery for larval and juvenile fish.

Mangroves are important coastal ecosystems for primary production, nutrient and sediment cycling and as a habitat provider. There are 4-5 important genera out of about 18, which are found in zones from sea to land, including *Sonneratia* found at the seaward side and is submerged at high tides; *Avicennia* on the seaward fringes; *Rhizophora* which may be found on the seaward fringe in Fiji or forming extensive forests in Thailand; *Bruguiera* which forms forests of tall trees at the landward edge and may merge into rainforest further inland.

They provide important habitats in the canopy for birds and insects and on the trunks and aerial roots as a hard substrate for snails and gastropods. Also found around roots are crustaceans such as the crab, *Uca* which use the soil surface to burrow as does the mudskipper. Barnacles attach to hard substrates. Also found are rock oysters and in water traps and rot holes, mosquitoes are found. In the western Atlantic, the small tidal range means that the roots of mangroves stay permanently in water and provide a habitat for sessile species such as sponges along with bryozoans and algae.

~~~~RESEARCH HIGHLIGHTS~~~~

Mangroves thrive on smothered coral reefs
On the islands of Sulawesi and Sumatra, in Indonesia, human induced change has smothered many of the fringing coral reefs and led to mangrove colonisation of the overlying soil. The rate of growth, as estimated from accumulation rates of forest biomass, were similar to other mangroves (1.5-8.1 tons/C/ha/yr). The study shows a clear shift from reef dominated to mangrove dominated habitats in parts of coastal Indonesia.
(Alongi et al. 2008)

Overall, the dominant species found in mangals appear to be crabs, while fish migrate in and out and use the area as a nursery for juveniles with an estimated 80% of commercially important fish having food chains linked back to mangals. Larger animals also use mangals such as turtles, crocodiles and alligators.

Fig 63. Mangroves. These mangrove trees are found along the shoreline at Lahaina, Maui in the Hawaiian Islands.

Mangals are important to indigenous peoples as they provide fish such as milkfish, molluscs and crabs as well as freshwater prawns such as *Macrobrachium*, which spawn in the mangals. They also increase productivity of coastal fisheries. Other products include honey, wood for building and thatching, medicines, fishing floats and chip wood (from *Rhizophora*).

Sediments and nutrients accumulate quickly, with runoff from land. And mangroves need to be conserved as many are being destroyed for industrial and residential development as well as in aid of tourism and other developments such as logging for charcoal and clearing for agriculture of sugar and rice. During the Vietnam War, somewhere in the region of 100,000 hectares of mangrove forest was destroyed with herbicides. Whether they will ever recover completely is unknown.

Mangroves are important coastal ecosystems for primary production, nutrient and sediment cycling and as a habitat provider.

~~~~RESEARCH HIGHLIGHTS~~~~

Mangroves increase the biomass of coral reef fish in the Caribbean
Mangrove forests are one of the world's most threatened tropical ecosystems with losses exceeding 35%. They provide an important habitat as nurseries to juvenile coral reef fish and may increase the survivorship of young fish by also serving as an intermediate nursery habitat. Mangroves in the Caribbean strongly influence the community structure of fish on neighbouring coral reefs, with the biomass of important species almost doubling when connected to mangroves. Mangroves deforestation is likely to have severe consequences for reef ecosystems.
(Mumby et al. 2004)

Mangals are being used detrimentally for aquaculture with 7% of Indonesian mangals cleared for tambacks or ponds to raise fish commercially. India has lost

40% of its mangals during the 20th century and much of Townsville on the Great Barrier Reef has been cleared for tourism and other development.

~~~~RESEARCH HIGHLIGHTS~~~~

The nutrient link between mangroves and coral reefs
Although mangrove forests have long been considered a source of organic matter for nearby reef systems, recent research has challenged the importance of this link and the quality of nutrient transfer. Using stable carbon isotope analysis, a recent study of reef consumers, including hard corals, sponges, a bivalve mollusc, tunicates and polychaetes shows that organic matter from mangroves is transferred in significant amounts to reef systems and accounts for up to 57% of the contribution. The source of the organic matter includes decaying mangrove leaves, microalgae, macroalgae and seagrass. The importance of this study is to show how strongly mangroves and coral reefs are linked and that both deserve our conservation management efforts.
(Granek et al. 2009)

Mangals are vital for primary production, fish, lobster and shrimp stocks and for prevention of sediment runoff from land into the sea. There are some measures being taken in areas such as Florida and Australia where boardwalks are being created through mangals that are important habitats for endangered species such as the American crocodile, brown pelican and the Atlantic Ridley sea turtle.

Conservation management

Coral reefs are used by commercial and sports fishermen, recreational snorkelers, divers and boat operators as well as researchers. There can be enormous pressure on certain reefs from human activities and this is predominantly because of the enormous numbers of people using reefs. Historically, coastal communities have fished reefs sustainably for thousands of years, using their bare hands, traps, lines, nets, harpoons and spears to catch fish and invertebrates on the reef. They used non-motorized boats and the small numbers of humans meant that there were few adverse impacts on the marine environment.

Mangals are vital for primary production, fish, lobster and shrimp stocks and for prevention of sediment runoff from land into the sea.

~~~~RESEARCH HIGHLIGHTS~~~~

Coral resilience in Discovery Bay
The structure of populations and the growth rates of corals from 2000 to 2008 in discovery Bay, North Jamaica, was studied for signs of resilience to multiple acute stressors of hurricanes and bleaching. The radical growth rates of branching corals at various reefs vary each but overall, the indication was that corals showed good levels of resistance on the fringing reefs around Discovery Bay.
(Crabbe 2009)

78

Since the 1940s, fish catches increased as the human population and the market for fish and other reef products increased. Nylon fishing nets were produced and fisheries became huge commercial affairs with bigger boats, refrigeration and a better transport network. Tourism and the trade in marine curios was also beginning to impact reefs. Now, as we have just seen in the previous section, the human impact on coral reefs worldwide is causing widespread destruction.

The management of coral reef fisheries generally involves designating certain areas for protection and restricting access as a short-term solution or creating marine protected areas as a more permanent solution. Fishers and fishing vessels are licensed, subject to catch limits and restricted to the use of certain fishing gear. Marine protected areas usually prioritise research, although tourism and recreational use can form an important part of conservation management, not least because it attracts vital funding.

Fig 64. Reef fisheries. The human impact of reef fishing is causing widespread destruction of reef ecosystems.

A major obstacle to successful management of marine reserve is that most of the world's coral reefs are found in poor, countries where marine conservation is not a priority. Management costs can be considerable and most protected tropical reefs are poorly managed and underfunded. Local communities do not always reap the benefits from jobs, tourism-businesses of based and so on, which often go to nonlocal people. Local communities can suffer economically where there are marine protected areas, fuelling illegal practices and corruption. Officials frequently turn a blind eye to illegal fishing practices. For many years there has been a

~~~~RESEARCH HIGHLIGHTS~~~~

Reserves protect reefs from predatory starfish
The crown-of-thorns starfish, *Acanthaster planci*, is a major management issue on coral reefs. It occurs throughout the Indian Pacific and shows waves of population outbreaks against a background of low densities. There have been three outbreaks affecting Australia's Great Barrier Reef since the 1960s, causing major destruction of coral across a large area -- dwarfing losses from other disturbances such as bleaching over the same period. Extensive surveys on the Great Barrier Reef Marine Park show that protection from fishing affects the frequency of outbreaks, with outbreaks 3.7 times higher on reefs that were open to fishing, than on no take reefs. Although exploited fishes are unlikely to prey on starfish directly, trophic cascades could favour invertebrates that pray on juvenile starfish, thereby reducing their numbers. (Sweatman 2008)

conflict of interest between local fishing communities and marine conservation managers. While marine parks and reserves were being set up after the Second World War, it is only recently that management strategies have really begun to embrace local communities in their action plans, although this remains challenging. Coral reefs

> Sustaining coral reef fisheries requires an integrated approach that includes the linking of social and ecological systems in any conservation management plan.

can be critical to the social and economic welfare of hundreds of millions of people, predominantly in developing countries. Sustaining coral reef fisheries requires an integrated approach that includes the linking of social and ecological systems in any conservation management plan.

Coral reefs are one of the ecosystems that are most vulnerable to climate change and while existing no-take marine protected areas are vital to conserve local reefs and their associated species, there needs to be more of them on a global scale.

Many marine biologists and conservationists believe that the management and conservation of coral reefs should continue to focus on conventional management tools such as establishing protected marine areas but should also be undertaking more active reef restoration. Projects such as the setting up of large-scale nurseries for rearing coral seedlings which are transplanted into damaged areas of nursery farmed coral colonies can help to repair extensive reef degradation (Rinkevich 2008).

Because all reefs are facing unprecedented pressures from the human impacts on the marine environment, the United Nations Environment Programme's World Conservation Monitoring Centre (UNEP-WCMC), along with the International Union for the Conservation of Nature (IUCN), it's part of the world's database on protected areas (WDPA). This is the most comprehensive global list of marine and terrestrial protected areas so far developed. It allows users to see information on marine protected areas and to visualise them in Google Earth. It is possible to download data and include other data such as species and ecosystem information. You can access this important new online system at www.wdpa-marine.org.

Fig 65. Conservation management. Effective management is essential to the protection of our coral reefs worldwide if further damage is to be avoided.

Reefs at risk

- **Pacific Ocean**

Australia
The Great Barrier Reef is protected by a marine park. About 20% of the reef is a no take zone, where fishing is off-limits. It is the largest reef system in the world and it is mostly in good condition. Australia's reefs are being well-managed and are at low risk, although they are still at risk locally from run-off of silt, nutrients and contaminants from agricultural urban and industrial areas. It was declared a UNESCO World Heritage Site in 1981.

Mining is banned but most of the area is open to fishing and diving as well as tourism development in some areas. Commercial prawn trawling is having an adverse effect on seafloor structure and biodiversity and is currently being investigated. Damage to the reef over the past 30 years has been from the crown-of-thorns starfish, *Acanthaster*, as well as a mass bleaching even in 1998. Generally, stakeholder involvement, educational programs and enforcement of achieving compliance for park regulations and management plans and the Great Barrier Reef world Heritage area is a classic example of a successful conservation programme (Bryant et al.1998).

Type: Barrier reef **Area:** 348,000 sq km **Location:** Queensland coast, northeastern Australia **Status:** Generally good but damaged by past bleaching event and crown-of-thorns starfish

The Great Barrier Reef is just 500,000 years old and the modern reef we see now only developed following the last ice age, 8,000 years ago. It is the largest coral reef system in the world – it is so large it can be seen from space and is the largest structure ever made by living organisms. It is a barrier reef with around 2,900 fringing and barrier reefs, continental islands, coral cays and 70 different bio-regions, making it the most ecologically diverse system in the world. There over 1,500 species of fish, 400 species of coral and 4,000 species of molluscs. Visitors to the reef include hammerhead and whale sharks.

Hawaii

The reefs around the Hawaiian islands comprise over 80% of coral reefs in the USA. Overall, species diversity is low but because they are so isolated from the corals in the rest of the world, many species have evolved in isolation and are endemic to the islands. Hawaiian reefs suffer predominantly from degradations related to urbanisation and development, including sewerage outfalls and run-off from tourist developments, causing sedimentation. The north western Hawaiian Islands and remote Pacific Islands, where there is little human impact remain relatively pristine. Many of Hawaii's coral reefs away from the heavily populated islands are in excellent condition and have been protected as National Wildlife Refuges (Rogers et al. 2002).

Type: Fringing reefs, atolls and submerged reefs **Area:** 2,000 sq km **Location:** North Pacific **Status:** Generally good but some degradation

The Hawaiian archipelago formed over millions of years from volcanic activity over hotspots in the Earth's mantle. As the Pacific plate moved northwest, volcanoes produced each island, a process that continues today. Fringing reefs are seen in the younger islands to the southeast of the chain, while atolls, such as Midway Atoll, are seen in the older islands in the northwest. They support some of the most abundant levels of marine species in the world, including over 700 species of fish, 400 species of algae and over 2,000 species of invertebrates.

- ### Indo-West Pacific

Indonesia

Human activities threaten over 85 percent of Indonesia's coral reefs, with nearly one half at high threat. The principal threats to Indonesian reefs are overfishing and destructive fishing, which threaten 64% and 53% of Indonesia's reefs.

Both coastal development and sedimentation from inland sources threaten about one fifth of the country's reefs. Indonesia's coral reefs help to support one of the largest marine fisheries in the world, generating 3.6 million tons of total marine fish production in 199 but corals and reef fish are endangered by destructive fishing practices. Cyanide and blast fishing are widespread, even in protected areas. Management is poor with few MPAs for corals covering some 6.2 million ha (Burke et al. 2002; Tun et al. 2004).

Type: Mostly fringing reefs **Area:** 51,000 sq km **Location:** South-east Asia **Status:** Variable with destruction from blasting

Indonesia has a coastline stretching over 95,000 km around more than 17,000 islands with some 51,000 km² of coral reefs and there may be others, as yet unmapped. It is thought that 51 percent of southeastern Asia's coral reefs and 18 percent of the world's coral reefs are found in Indonesian waters. There are more than 480 species of hard coral and more than 1,650 species of fish. The extent of Indonesia's biological endowment is still unknown.

Philippines

High population numbers and poverty have resulted in overfishing in over 80% of reef areas especially of demersal fish stocks. Destructive fishing methods are also a major concern with blast fishing and the use of poisons. *Muro-ami*, is the practice of sending a line of divers to depths of 10-30 m with metal weights. They then bang on corals to drive fish out into nets. This has proved extremely damaging to reefs and has been banned since 1986 but illegal practices continue in new, more remote areas. Destructive fishing has destroyed 70% of the fisheries within 15 km of the shore. Coastal development, land clearance, agriculture and aquaculture also threatens reefs, as are untreated waste discharges and some 35-40% of reefs are threatened. In all, it is estimated that 98% of Philippine reefs are threatened by human activities (70% at high risk).

The Philippine government is not enforcing effective management strategies, although there have been some successful local management projects(Burke et al. 2002).

Type: Area: 26,000 sq km **Location:** South-east Asia **Status:** Highly threatened

The coral reef area at the Philippines is the second largest in southeastern Asia, at 26,000 km^2, and has high species diversity. There are some 400 species of stony coral (12 endemic) and 915 reef fish species.

- **Atlantic Ocean**

Bahamas
Some of the least threatened coral reefs in the Caribbean region are found in the Bahamas and Turks and Caicos Islands, with only about 30% of coral reefs in the area identified as threatened by overfishing -- the only threat identified in most areas. There is some threat from coastal development and pollution in some of the more heavily populated islands but the reefs are in good condition. The growth in tourism has led to destruction of coastal habitats future hotel and marina development as well as damage to corals by recreational divers. Tourism has also led to problems such as waste management.

Grouper and conch are showing evidence of overfishing but reef fishes do not appear to be overexploited, although there are concerns about poaching by foreign fishers using illegal methods, mostly from Haiti and the Dominican Republic. The government of the Bahamas pioneered a reef protection by establishing its first Land and Sea National Park in 1958 in Exuma Cays and in 1986 he became the first Caribbean area to make the park into a no-take fishery. Ten new national parks were established in 2002 (Burke and Maidens 2004).

Type: Fringing reefs, patch reefs, barrier reefs. **Area:** 3,150 sq km **Location:** Little Bahama and Great Bahama Banks, Bahamas, West Atlantic **Status:** Generally good

The Bahamas consists of two limestone platforms around 4,500 m thick, with an archipelago of 700 islands, many with fringing reefs. The platforms form the Little Bahama and Great Bahama Banks. The coral platforms have been accumulating for some 70 million years and corals are found around 10-25 m below the sea surface.

Belize

The most pervasive threat for the reefs of Belize is overfishing by both industrial fishing fleets and small-scale local fisheries. Larger cays and tourist areas such as Amergis Caye and San Pedro Town of growing rapidly, causing problems with sediment. Off southern Belize and continental Honduras, run-off from intensification of agriculture and logging has caused problems for reef systems over the last few decades. Fertiliser run-off from banana and citrus plantations is also causing nutrient pollution from southern Belize down through Guatemala and Honduras, although initiative such as the Better Banana Project are addressing these problems.

In 1998, high sea-surface temperatures followed by Hurricane Mitch (a category 5 storm) caused bleaching and severe damage and coral loss in the lagoonal reefs of Belize and coral destruction in the fore reefs and outer atoll reefs. We still do not know the full consequences of this natural disturbance. Belize has a system of 13 marine protected areas and has a well-established management strategy under the Belize Coastal Zone Management Authority and Institute (Burke and Maidens 2004). It was declared a World Heritage Site in 1996.

Type: Barrier with atolls and patch reefs **Area:** 300 km **Location:** Western Caribbean **Status:** Good

Belize is a 300 km section of the much larger Mesoamerican Barrier Reef system, which is around 960 km long – the second largest reef system to the Great Barrier Reef. The area includes Glover's Reef Marine Reserve, Great Blue Hole, Half Moon Caye, Hol Chan Marine Reserve. There are 450 cays. It supports a large diversity of species, including 70 species of stony coral, 36 soft coral species, 500 species of fish. There are likely to be many species as yet undiscovered in this little researched reef system.

Bermuda

Bermuda's reefs are the most northerly coral reefs in the world. They survive because of the warm eddies of the Gulf Stream, and although the species diversity is less than that seen in the more southerly Caribbean reefs, they are an important ecosystem, with some 20 species of stony coral and 17 species of soft coral. There are also around 120 species of fish associated with the reefs.

All of Bermuda's reefs are predominantly threatened from overfishing, although 60% of reefs are threatened from tourism-based activities, particularly cruise ships. Coastal development also threatens some 50% of reefs. Declines have been seen since the early 1990s.

However, overall, Bermuda's reefs are healthy and free from disease and bleaching (Burke and Maidens 2004).

Type: Atolls, fringing and patch reefs **Area:** 370 sq km **Location:** Northwest Atlantic **Status:** Generally good

The Bermuda Platform developed from a seamount, or submarine volcano, the surface of which now lies some 15-18m below sea level. Coral reefs have developed and grown upward over millions of years to form a thick layer of limestone, which also forms the islands of Bermuda to the south-east. The coral growth forms an atoll with patch reefs at the centre.

Cayman Islands

The main problem with the Cayman Island reefs, despite being managed under strict marine conservation laws and his own system of NPAs, is overfishing of conch and lobster. Intensive tourism, including impacts from cruise ships, and coastal development are also a major concern. A recent analysis found around 80% of reefs are threatened, although most reefs are generally in good condition. The main area of impact is around the more developed island of Grand Cayman, which is the focus of the dive industry (Burke and Maidens 2004).

Type: Atolls, fringing and patch reefs **Area:** 370 sq km **Location:** Northwest Atlantic **Status:** Generally good

The Cayman Islands are found in the centre of the Caribbean Sea (part of the Greater Antilles: Cuba, the Cayman Islands, Jamaica, Haiti and the Dominican Republic), and their coral reefs cover over 8,600 km².

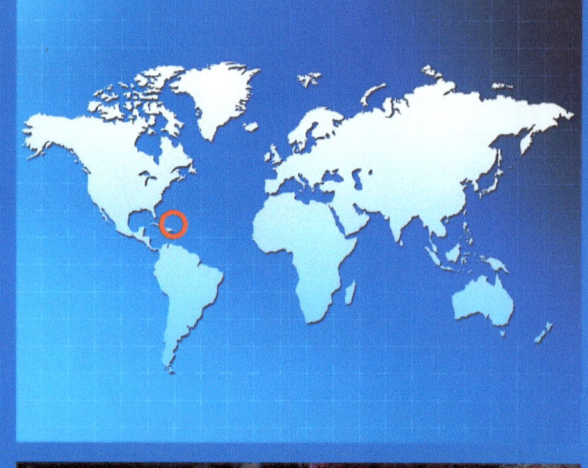# TROPICAL MARINE ENVIRONMENTS

Dominican Republic
Over 80% of these reefs are threatened from multiple sources including human activities with one third under very high threat. The Dominican Republic is a poor tropical country with widespread unemployment, densely populated coastal zones, easy access to the reefs and a narrow shelf area. Island nations also rely heavily on agriculture and export of sugar, coffee, bananas and tobacco and this has created problems for reefs due to land clearance and increased erosion, which caused sedimentation. All this means that the reef resources have been heavily used by the local community, with illegal fishing activities frequent and a lack of enforcement of regulations, despite international trade in conch from the Dominican Republic being banned under CITES. The growth in tourism has alleviated some of the social and economic problems of the area, but the unmanaged coastal development is currently threatening some 70% of reefs (Burke and Maidens 2004).

Type: Fringing and barrier reefs **Area:** 8,600 sq km **Location:** Northwest Atlantic **Status:** Poor

One of the Greater Antilles, the coral reefs of the Dominican Republic cover over 8,600 km² with some important offshore bank reefs, including Silver Bank, a major breeding area for North Atlantic humpback whales.

- **Indian Ocean**

Maldives

A severe bleaching event in 1998 killed up to 90% of the corals in some areas of the Maldives, with some, but not complete, recovery several years later. The southernmost atolls were least affected by bleaching. The tourist industry has increased enormously since the 1970s and there are the usual threats from human disturbance. However, this does not appear to be impacting on the reefs at present. The reef fishery is thought to be sustainable with relatively high fish abundance and generally, apart from the past damage from bleaching, the reef is healthy.

In order to stop reef fishing, the Maldivian government set up 10 MPAs in 1995 and a further 10 in 1999. However, with the subsequent lack of government support, enforcement is weak and resources are few to implement management strategies.

Type: Fringing reefs, atolls and patch reefs **Area:** 9,000 sq km **Location:** Western Indian Ocean **Status:** Reasonable with good recovery from bleaching

The Maldives consists of 26 atolls, many of which are large and arranged in an elliptical ring around 800 km long. Many atolls have a separate reefs and coralline islets. Patch reefs and faros (like mini atolls, only found in the Maldives) are found within the atoll lagoons. There are more than 200 stony corals, and over 1,000 fish species.

Red Sea

Tourism has taken its toll on localised reefs, especially around Egypt, with it's world famous dive sites. The coastal areas of the Gulf of Aqaba in Jordan, the Egyptian coast and industrial areas of the Gulf of Suez, Yanbu and Jeddah in Saudi Arabia are the major concerns. Parts of the reef have suffered from predation from Crown-of-thorns starfish, *Acanthaster*. The Red Sea is an important shipping route and oil spills are also a major threat as tankers make their way to the Suez Canal. An estimated 100 million tons of oil is transported annually by 20,000-35,000 tankers. Fish are collected for the aquarium trade in Saudi Arabia and Yemen, with at least seven fish exporters working in the area.

The Egyptian government set up the Ras Mohammed National Park in 1983 because of the rich species diversity of the area.

Type: Barrier, fringing, atolls and patch reefs **Area:** 16,500 sq km **Location:** Red Sea coasts **Status:** Localised damage but generally good

The Red Sea was formed when Arabia split from Africa during tectonic movements which began in the Eocene and continue today. It is a diverse ecosystem with 2,000 km of fringing coral reef, some 5,000-7,000 years old. Corals are found mainly along the northern and central coasts and there are more than 250 corals and 1,000 invertebrates with over 1,200 species of fish, many endemic to the Red Sea. Visitors to the reef include 44 species of shark.

Study questions:

1. Discuss the structure of corals and their association with symbiotic zooxanthellae
2. Where are corals found in the world and what are their limiting growth factors?
3. What is being done to conserve coral reef environments?
4. Describe how coral atolls form and discuss zonation on the reef
5. Discuss the various species associated with coral reefs
6. Discuss symbiotic relationships between reef animals
7. Discuss the threats corals face
8. Which are the most threatened reefs in the world?

Summary

- Coral reefs are one of the most diverse marine habitats on Earth with primary productivity between 30 and 250 times that of the open ocean.

- Most reef building corals have a symbiotic relationship with dinoflagellates known as zooxanthellae, members of the phytoplankton.

- There are six major physical factors that limit coral reef development: temperature, depth, light, salinity, sedimentation and emersion into air.

- Coral reefs are found in three principal areas; the Caribbean, including the Bahamas and Florida Keys, The Red Sea, including Indian Ocean Islands such as the Seychelles and the Indo West Pacific. The centre of diversity is in the Indo West Pacific

- Corals are generally found in a broad band throughout the tropics, within the 20^0C isotherms with extensions where there are warm water currents and are the result of growth since the last ice age,10-11,000 yrs ago, although they have much older foundations, particularly the atolls.

- There are 4 types of reef: fringing reefs, barrier reefs, patch reefs and atolls. All are part of a series of forms which all develop in the same way.

- Coral reefs cover approximately 600,000 km^2, around 0.2% of the global ocean and about 15% of the shallow sea to a depth of about 30 m. The Great Barrier Reef is the largest, covering an area around 2,000 km by 145 km.

- Corals can only cope with increasing sea levels at a rate of approximately 4.5 cm per decade.

- Associated species are also rich in diversity on reefs with around 500 species of reef fish in the Bahamas, 1,500 in the Great Barrier Reef and 2,000 in the Philippines.

- Representatives of almost all the phyla are found within the coral reef ecosystem.

- Some 25% of the marine fishes in the world are found only in reefs. Most fish are specialized feeders and are important to the ecology of the reef.

- There are many examples of symbiotic relationships between animals on the reef, the most widespread being the relationship between coral polyps and zooxanthellae, clownfishes which live in the stinging nematocysts of anemones, remoras and the cleaning symbiosis of cleaner wrasse.

- Threats to coral reefs include physical or biological factors, either directly by hurricanes, predators (such as *Acanthaster*) or human disturbance, for example, or indirectly by overfishing or climate warming.

- Bleaching events occur when zooxanthellae leave their host and can occur when temperatures rise acutely by 3-4^{0}C or during a prolonged rise of 1-2^{0}C. The severity and frequency of bleaching events has increased since the 1960's in many important species of reef building coral such as *Acropora palmata, Posillopora, Monastrea, Millepora* and *Helipora*.

References

A.S Grutter, Effect of the removal of cleaner fish on the abundance and species composition of reef fish, *Oecologia* 111 (1997), pp. 137–143.

A.S Grutter, Parasite removal rates by the cleaner wrasse *Labroides dimidiatus*, *Mar. Ecol. Prog. Ser.* **130** (1996), pp. 61–70.

A.S Grutter, Spatio-temporal variation and feeding selectivity in the diet of the cleaner fish *Labroides dimidiatus*, *Copeia* 1997 (1997), pp. 346–355.

A.S Grutter, The relationship between cleaning rates and ectoparasite loads in coral reef fishes, *Mar. Ecol. Prog. Ser.* 118 (1995), pp. 51–58.

Alexandra S Grutte, Jan Maree Murphy and J.Howard Choat (2003) Cleaner Fish Drives Local Fish Diversity on Coral Reefs. *Current Biology.* Vol 13, Issue 1. P64-76

Anthony, K. R. N.; Kline, D. I.; Diaz-Pulido, G.; Dove, S.; Hoegh-Guldberg, O. (2008) **Ocean acidification causes bleaching and productivity loss in coral reef builders.** Proceedings of the National Academy of Sciences of the United States of America 105(45), 17442-17446

Antonius, A. (1973). **"New observations on coral destruction in reefs.".** 10th Meeting Assoc. 1st. Mar. Lab. Carib. **10** (3 (abstract)).

Aronson RB, Precht WF. White-band disease and the changing face of Caribbean coral reefs. *Hydrobiologia.* 2001;460:25–38.

B. J. Godley, A. C. Broderick, F. Glen, G. C. Hays (2002) **Temperature-dependent sex determination of Ascension Island green turtles**. Mar Ecol Prog. Ser Vol. 226: 115–124

B. Riegl, S. J. Purkis, J. Keck, G. P. Rowlands (2009). **Monitored and modeled coral population dynamics and the refuge concept** Marine Pollution Bulletin 58(1), 24-38

B. Rinkevich, Baruch (2008). **Management of coral reefs: We have gone wrong when neglecting active reef restoration** Marine Pollution Bulletin 56(11), 1821-1824

Baker, P. E. (1992) **Oceanic islands and the mantle: historical perspectives.** Journal of Volcanology and Geothermal Research 50(1-2), 17-32.

Baums IB, Miller MW, Hellberg ME. Regionally isolated populations of an imperiled Caribbean coral, *Acropora palmata*. *Molecular Ecology.* 2005;14:1377–1390.

Benjamin S. Halpern; Fiorenza Micheli; Helen E. Fox; Matthew T. Perry (2008) **A Global Map of Human Impact on Marine Ecosystems** SCIENCE VOL 319

Bruno, John F.; Selig, Elizabeth R.; Casey, Kenneth S.; Page, Cathie A.; Willis, Bette L.; Harvell, C. Drew; Sweatman, Hugh; Melendy, Amy M. (2007) **Thermal stress and coral cover as drivers of coral disease outbreaks.** PLoS Biology 5(6), 1220-1227.

Bryant, D., L. Burke, J. McManus and M. Spalding (1998) **Reefs at Risk: A map-based indicator of threats to the world's coral reefs**. World Resources Institute

Burke, L. and J. Maidens (2004) **Reefs at Risk in the Caribbean.** World Resources Institute, Washington.

Burke, L., E. Selig and M. Spalding (2002) **Reefs At Risk in Southeast Asia.** World Resources Institute

C. Scheibner, R. P. Speijer (2008). **Decline of coral reefs during late Paleocene to early Eocene global warming** Universitaet Bremen, Bremen, Germany. eEarth (2008), 3(1), 19-26

Callum M. Roberts, Colin J. McClean, John E. N. Veron et al. (2002) Marine Biodiversity Hotspots and Conservation Priorities for Tropical Reefs. Science 15th February. Vol. 295. No.5558. pp. 1280-1284

Carpenter, Kent E.; Abrar, Muhammad; Aeby, Greta; Aronson, Richard B.; Banks, Stuart; Bruckner, Andrew; Chiriboga, Angel; et al. (2008) **One-Third of Reef-Building Corals Face Elevated Extinction Risk from Climate Change and Local Impacts.** Science (Washington, DC, United States) 321(5888), 560-563.

Casas V, Kline DI, Wegley L, Yu YN, Breitbart M, et al. Widespread association of a *Rickettsiales*-like bacterium with reef-building corals. *Environmental Microbiology.* 2004;6:1137–1148.

Cervino, J. M.; Hayes, R. L.; Honovich, M.; Goreau, T. J.; Jones, S.; Rubec, P. J. (2003) **Changes in zooxanthellae density, morphology, and mitotic index in hermatypic corals and anemones exposed to cyanide.** Marine Pollution Bulletin 46(5), 573-586.

Chen, Tian-ran; Yu, Ke-fu; Lin, Zhi-fen; Chen, Te-gu. (2006) **The advances in the study of the response of coral reefs to red tides and the possible record of historical red tides in reef corals.** Yanshi Kuangwuxue Zazhi 25(6), 523-529.

D. Bryant, L. Burke, J. McManus, M. Spalding (1998) *Reefs at Risk: A Map-Based Indicator of Potential Threats to the World's Coral Reefs* (World Resources Institute, Washington, DC;

International Center for Living Aquatic Resource Management, Manila; and United Nations Environment Programme-World Conservation Monitoring Centre, Cambridge, 1998).

D. M. Alongi, L. A. Trott, R. Rachmansyah, F. Tirendi, A. D. McKinnon, M. C. Undu (2008) **Growth and development of mangrove forests overlying smothered coral reefs, Sulawesi and Sumatra, Indonesia** Marine Ecology: Progress Series 370 97-109

D. Steinke, Dirk; T. S. Zemlak, P. D. N. Hebert (2009) **Barcoding nemo: DNA-based identifications for the ornamental fish trade.** Canadian Centre for DNA Barcoding, Biodiversity Institute of Ontario, University of Guelph, Guelph, ON, Can. PLoS One 4(7),

Dameron, Oliver J.; Parke, Michael; Albins, Mark A.; Brainard, Russell. (2007) **Marine debris accumulation in the Northwestern Hawaiian Islands: An examination of rates and processes.** Marine Pollution Bulletin 54(4), 423-433.

Danovaro, Roberto; Bongiorni, Lucia; Corinaldesi, Cinzia; Giovannelli, Donato; Damiani, Elisabetta; Astolfi, Paola; Greci, Lucedio; Pusceddu, Antonio (2008) **Sunscreens cause coral bleaching by promoting viral infections.** Environmental Health Perspectives 116(4), 441-447.

De Goeij, Jasper M.; van Duyl, Fleur C. (2007) **Coral cavities are sinks of dissolved organic carbon (DOC)** Limnology and Oceanography 52(6), 2608-2617.

DeBose, Jennifer L.; Lema, Sean C.; Nevitt, Gabrielle A. (2008). **Dimethylsulfoniopropionate as a Foraging Cue for Reef Fishes.** Science (Washington, DC, United States) 319(5868), 1356.

Downs, Craig A.; Kramarsky-Winter, Esti; Martinez, Jon; Kushmaro, Ariel; Woodley, Cheryl M.; Loya, Yossi; Ostrander, Gary K. (2009) **Symbiophagy as a cellular mechanism for coral bleaching.** Autophagy (2009), 5(2), 211-216.

Dulvy, Nicholas K. (2006) **Conservation Biology: Strict Marine Protected Areas Prevent Reef Shark Declines.** Current Biology 16(23), R989-R991.

E. F. Granek, J. E. Compton, D. L. Phillips (2009) **Mangrove-Exported Nutrient Incorporation by Sessile Coral Reef Invertebrates** Ecosystems 12(3), 462-472.

E. Kramarsky-Winter, Y. Loya, M. Vizel, C. A. Downs (2009) **Method for coral tissue cultivation and propagation** PCT Int. Appl. (2009), Patent written in English. Application: WO 2008-IL1236 20080917. Priority: US 2007-973061 20070917.

E. Wolanski, J. A. Martinez, R. H. Richmond (2009). **Quantifying the impact of watershed urbanization on a coral reef: Maunalua Bay, Hawaii** Estuarine, Coastal and Shelf Science 84(2), 259-268

Eckes, Maxi J.; Siebeck, Ulrike E.; Dove, Sophie; Grutter, Alexandra S. (2008) **Ultraviolet sunscreens in reef fish mucus.** Marine Ecology: Progress Series 353 203-211.

Edwards, Alasdair J.; Clark, Susan. (1999) **Coral transplantation: a useful management tool or misguided meddling?** Marine Pollution Bulletin Vol (8-12), 474-487.

Froukh, Tawfiq; Kochzius, Marc. (2007) **Genetic population structure of the endemic fourline wrasse (Larabicus quadrilineatus) suggests limited larval dispersal distances in the Red Sea.** Molecular Ecology 16(7), 1359-1367.

G. Wei, M. T. McCulloch, G. Mortimer, W. Deng, L. Xie (2009). **Evidence for ocean acidification in the Great Barrier Reef of Australia** Geochimica et Cosmochimica Acta 73(8), 2332-2346

Gardner TA, Cote IM, Gill JA, Grant A, Watkinson AR. Long-term region-wide declines in Caribbean corals. *Science.* 2003;301:958–960.

Gardner, Toby A.; Cote, Isabelle M.; Gill, Jennifer A.; Grant, Alastair; Watkinson, Andrew R. (2003) **Long-Term Region-Wide Declines in Caribbean Corals.** Science (Washington, DC, United States) 301(5635), 958-961.

Gerlach, Gabriele; Atema, Jelle; Kingsford, Michael J.; Black, Kerry P.; Miller-Sims, Vanessa. (2007) **Smelling home can prevent dispersal of reef fish larvae** Proceedings of the National Academy of Sciences of the United States of America 104(3), 858-863.

Gleason, D. F., and Welligton, G. M. (1993). **Ultraviolet radiation and coral bleaching.** Nature, 365, 836-838.

Glud, Ronnie N.; Eyre, Bradley D.; Patten, Nicole (2008) **Biogeochemical responses to mass coral spawning at the Great Barrier Reef: effects on respiration and primary production.** Limnology and Oceanography 53(3), 1014-1024.

Godley BJ, Broderick AC, Glen F, **Hays GC** (2002). Temperature dependent sex determination of Ascension Island green turtles. *Marine Ecology Progress Series* 226, 115-124.

Grutter, Alexandra S.; Murphy, Jan Maree; Choat, J. Howard. (2003) **Cleaner Fish Drives Local Fish Diversity on Coral Reefs** Current Biology 13(1), 64-67.

H. Hasler, J.A. Ott (2008). **Diving down the reefs? Intensive diving tourism threatens the reefs of the northern Red Sea.** Marine Pollution Bulletin (2008), 56(10), 1788-1794

H. M. Guzman, R. Cipriani, J. B. C. Jackson (2008) **Historical decline in coral reef growth after the Panama Canal** Ambio 37(5), 342-346

Hughes TP. Catastrophes, Phase-Shifts, and Large-Scale Degradation of a Caribbean Coral-Reef. *Science.* 1994;265:1547–1551.

Hughes, Terence P.; Rodrigues, Maria J.; Bellwood, David R.; Ceccarelli, Daniela; Hoegh-Guldberg, Ove; McCook, Laurence; Moltschaniwskyj, Natalie; Pratchett, Morgan S.; Steneck, Robert S.; Willis, Bette. (2007) **Phase Shifts, Herbivory, and the Resilience of Coral Reefs to Climate Change.** Current Biology 17(4), 360-365.

J. B. Ries, S. M. Stanley, L. A. Hardie, (2006) Scleractinian corals produce calcite, and grow more slowly, in artificial Cretaceous seawater Geology 34, 525

J. E. Cinner, T. R. McClanahan, T. M. Daw, N. A. J. Graham, J. Maina, S. K. Wison, T. P. Hughes (2009). **Linking Social and Ecological Systems to Sustain Coral Reef Fisheries** Current Biology 19(3), 206-212

J. Lian, H. Huang, Y. Li, Z. Dong, J. Yang, G. Zhou, F. You (2009) **Method for restoring coral reef ecosystem and device for reef building coral larva to attach** Faming Zhuanli Shenqing Gongkai Shuomingshu, CODEN: CNXXEV CN 101406168 A 20090415 Patent written in Chinese. Application: CN 2008-10219217 20081118

J. Mallela, M. J. C. Crabbe (2009) **Hurricanes and coral bleaching linked to changes in coral recruitment in Tobago** Marine Environmental Research 68(4), 158-162.

J. Silverman, B. Lazar, L. Cao, K. Caldeira, J. Erez (2009) **Coral reefs may start dissolving when atmospheric CO2 doubles** Geophysical Research Letters (2009), 36(5), L05606/1-L05606/5.

Jackson, Jeremy B. C. (2008) **Ecological extinction and evolution in the brave new ocean.** Proceedings of the National Academy of Sciences of the United States of America 105(Suppl. 1), 11458-11465.

Jarosław Stolarski, et al. (2007) A Cretaceous Scleractinian Coral with a Calcitic Skeleton. Science Oct 5th Vol. 318. No 5847, pp. 92-94

John M. Pandolfi et al. (2003) Global Trajectories of the Long-Term Decline of Coral Reef Ecosystems. Science vol. 301.no 5635, pp955-958

Johnson, Kenneth G.; Jackson, Jeremy B. C.; Budd, Ann F. (2008) **Caribbean Reef Development Was Independent of Coral Diversity over 28 Million Years.** Science (Washington, DC, United States) 319(5869), 1521-1523.

Jones, G. P.; Milicich, M. J.; Emslie, M. J.; Lunow, C. (1999) **Self-recruitment in a coral reef fish population** Nature (London) 402(6763), 802-804.

Jones, Ross J.; Steven, Andrew L. (1997) **Effects of cyanide on corals in relation to cyanide fishing on reefs.** Marine and Freshwater Research 48(6), 517-522.

K. L. Cheney, A. S. Grutter, S. P. Blomberg, N. J. Marshall (2009) **Blue and Yellow Signal Cleaning Behavior in Coral Reef Fishes.** Current Biology 19(15), 1283-1287.

Kelman, Dovi; Kashman, Yoel; Rosenberg, Eugene; Kushmaro, Ariel; Loya, Yossi. (2006) **Antimicrobial activity of Red Sea corals.** Marine Biology (Heidelberg, Germany) 149(2), 357-363.

Kiessling W. 2001a. Paleoclimatic significance of Phanerozoic reefs. *Geology* 29:751–54

Knowlton N, Lang JC, Keller BD. Case study of natural population collapse: post-hurricane predation of Jamaican staghorn corals. *Smithsonian Contributions Marine Science.* 1990;31:1–25.

Kruse PD, Zhuravlev AY, James NP. 1995. Primordial metazoan-calcimicrobial reefs: Tommotian (Early Cambrian) of the Siberian Platform. *Palaios* 10:291–321

Kuffner, Ilsa B.; Andersson, Andreas J.; Jokiel, Paul L.; Rodgers, Ku ulei S.; MacKenzie, Fred T. (2008) **Decreased abundance of crustose coralline algae due to ocean acidification.** Nature Geoscience 1(2), 114-117.

Kuo, Yau-Lun; Yu, Gwo-Lin; Yang, Yeh-Lin; Wang, Hsiang-Hua (2007) **Effects of typhoon disturbances on understory light and seedling growth of six tree species in a forest at Kenting, southern Taiwan.** Taiwan Linye Kexue 22(4), 367-380.

L. Cao, K. Caldeira (2008) **Atmospheric CO2 stabilization and ocean acidification** Geophysical Research Letters (2008), 35(19), L19609/1-L19609/5.

Lecchini, David; Polti, Sandrine; Nakamura, Yohei; Mosconi, Pascal; Tsuchiya, Makoto; Remoissenet, Georges; Planes, Serge (2006) **New perspectives on aquarium fish trade.** Fisheries Science (Carlton, Australia) 72(1), 40-47.

Lirman, Diego; Fong, Peggy. (2007) **Is proximity to land-based sources of coral stressors an appropriate measure of risk to coral reefs? An example from the Florida Reef Tract.** Marine Pollution Bulletin 54(6), 779-791.

Lobel PS (2008) **Diver Eco-Tourism and the Behavior of Reef Sharks and Rays – an Overview**. In: Brueggeman P, Pollock NW, eds. Diving for Science 2008. Proceedings of the American Academy of Underwater Sciences 27th Symposium. Dauphin Island, AL: AAUS; 2008.

Lough, Janice M. (2008) **10th Anniversary Review: a changing climate for coral reefs** Journal of Environmental Monitoring 10(1), 21-29.

M. J. Atkinson, P. Cuet (2008) **Possible effects of ocean acidification on coral reef biogeochemistry: topics for research** Marine Ecology: Progress Series 373 249-256

M. J. C. Crabbe (2009). **Scleractinian coral population size structures and growth rates indicate coral resilience on the fringing reefs of North Jamaica** Marine Environmental Research 67(4-5), 189-198

M. J. Paddack, D. J. Reynolds, C. Aguilar, R. S. Appeldoorn, J. Beets, E. W. Burkett, P. M. Chittaro, K. Clarke, R. Esteves, A. C. Fonseca, G. E. Forrester, A. M. Friedlander, J. Garcia-Sais, G. Gonzalez-Sanson, L. K. B. Jordan, D. B. McClellan, M. W. Miller, P. P. Molloy, R. J. Mumby, I. Nagelkerken, M. Nemeth, R. Navas-Camacho, J. Pitt, N. V. C. Polunin, M. C. Reyes-Nivia, D. R. Robertson, A. Rodriguez-Ramirez, E. Salas, S. R. Smith, R. E. Spieler, M. A. Steele, I. D. Williams, C. L. Wormald, A. R. Watkinson, I. M. Cote, Isabelle (2009). **Recent Region-wide Declines in Caribbean Reef Fish Abundance** Current Biology 19(7), 590-595.

Madkour, Hashem Abbas; Dar, Mahmoud A. (2007) **The anthropogenic effluents of the human activities on the Red Sea coast at Hurghada Harbour (case study).** Egyptian Journal of Aquatic Research 33(1), 43-58.

Mak, Karen K. W.; Yanase, Hideshi; Renneberg, Reinhard. (2005) **Cyanide fishing and cyanide detection in coral reef fish using chemical tests and biosensors.** Biosensors & Bioelectronics 20(12), 2581-2593.

Mangi, S. C.; Roberts, C. M. (2006) **Quantifying the environmental impacts of artisanal fishing gear on Kenya's coral reef ecosystems.** Marine Pollution Bulletin 52(12), 1646-1660.

Manzello, Derek P.; Brandt, Marilyn; Smith, Tyler B.; Lirman, Diego; Hendee, James C.; Nemeth, Richard S. (2007) **Hurricanes benefit bleached corals** Proceedings of the National Academy of Sciences of the United States of America 104(29), 12035-12039.

Marhaver, Kristen L.; Edwards, Robert A.; Rohwer, Forest (2008) **Viral communities associated with healthy and bleaching corals.** Environmental Microbiology 10(9), 2277-2286.

Marion, Guy S.; Dunbar, Robert B.; Mucciarone, David A.; Kremer, James N.; Lansing, J. Stephen; Arthawiguna, Alit. (2005) **Coral skeletal 15N reveals isotopic traces of an agricultural revolution.** Marine Pollution Bulletin 50(9), 931-944.

McCulloch, M. T.; Esat, T. (2000) **The coral record of last interglacial sea levels and sea surface temperatures.** Chemical Geology 169(1-2), 107-129.

McNeil, Ben I.; Matear, Richard J. (2008) **Southern Ocean acidification: a tipping point at 450-ppm atmospheric CO2.** Proceedings of the National Academy of Sciences of the United States of America 105(48),18860-18864

Mehta, Rita S.; Wainwright, Peter C. (2007) **Raptorial jaws in the throat help moray eels swallow large prey.** Nature (London, United Kingdom) 449(7158), 79-82.

Mumby, Peter J.; Edwards, Alasdair J.; Ernesto Arias-Gonzalez, J.; Lindeman, Kenyon C.; Blackwell, Paul G.; Gall, Angela; Gorczynska, Malgosia I.; Harborne, Alastair R.; Pescod, Claire L.; Renken, Henk; C. C. Wabnitz, Colette; Llewellyn, Ghislane. (2004) **Mangroves enhance the biomass of coral reef fish communities in the Caribbean.** Nature (London, United Kingdom) 427(6974), 533-536.

N. A. J. Graham, T. R. McClanahan, M. A. MacNeil, S. K. Wilson, N. V. C. Polunin, S. Jennings, P. Chabanet, S. Clark, M. D. Spalding, Y. Letourneur, L. Bigot, R. Galzin, M. C. Ohman, K. C. Garpe, A. J. Edwards, C. R. C. Sheppard (2008) **Climate warming, marine protected areas and the Ocean-scale integrity of coral reef ecosystems.** PLoS One 3(8). School of Marine Science & Technology, Newcastle University, Newcastle-upon-Tyne, UK.

Nakamura, Masaru; Kobayashi, Yasuhisa; Miura, Saori; Alam, Mohamad Ashraful; Bhandari, Ramji Kumar. (2006) **Sex change in coral reef fish.** Fish Physiology and Biochemistry Volume Date 2005, 31(2-3), 117-122.

Newton, Katie; Cote, Isabelle M.; Pilling, Graham M.; Jennings, Simon; Dulvy, Nicholas K. (2007) **Current and Future Sustainability of Island Coral Reef Fisheries.** Current Biology 17(7), 655-658.

Nott, Jonathan; Hayne, Matthew. (2001) **High frequency of 'super-cyclones' along the great barrier reef over the past 5,000 years.** Nature (London, United Kingdom) 413(6855), 508-512. Knowlton, Nancy (2001) **The future of coral reefs.** Proceedings of the National Academy of Sciences of the United States of America 98(10), 5419-5425.

P. L. Munday, D. L. Dixson, J. M. Donelson, G. P. Jones, M. S. Pratchett, G. V. Devitsina, K. B. Doving (2009) **Ocean acidification impairs olfactory discrimination and homing ability of a marine fish.** Proceedings of the National Academy of Sciences of the United States of America 106(6), 1848-1852.

P.F. Cowman, D. R. Bellwood, L. van Herwerden (2009). **Dating the evolutionary origins of wrasse lineages (Labridae) and the rise of trophic novelty on coral reefs** Molecular Phylogenetics and Evolution 52(3), 621-631

Perkol-Finkel, S.; Shashar, N.; Benayahu, Y. (2006) **Can artificial reefs mimic natural reef communities? The roles of structural features and age.** Marine Environmental Research 61(2), 121-135.

Pratchett, M. S.; Schenk, T. J.; Baine, M.; Syms, C.; Baird, A. H (2009) **Selective coral mortality associated with outbreaks of Acanthaster planci L. in Bootless Bay, Papua New Guinea.** Marine Environmental Research 67(4-5), 230-236.

R. Packett, C. Dougall, K. Rohde, R. Noble (2009) **Agricultural lands are hot-spots for annual runoff polluting the southern Great Barrier Reef lagoon** Marine Pollution Bulletin 58(7), 976-986

R. S. Calhoun, M. E. Field (2008) **Sand composition and transport history on a fringing coral reef, Molokai, Hawaii** Journal of Coastal Research 24(5), 1151-1160

R. W. Buddemeier, P. L. Jokiel, K. M. Zimmerman, D. R. Lane, J. M. Carey, G. C. Bohling, J. A. Martinich (2008), **A modeling tool to evaluate regional coral reef responses to changes in climate and ocean chemistry.** American Society of Limnology and Oceanography. 6(Sept.) 395-411

R. P. van Dam, C. F. Diez (1996) **Diving behavior of immature hawksbills (*Eretmochelys imbricata*) in a Caribbean cliff-wall habitat** Marine Biology Volume 127, Number 1

Raymundo, L. J.; Maypa, A. P.; Gomez, E. D.; Cadiz, Pablina. (2007) **Can dynamite-blasted reefs recover? A novel, low-tech approach to stimulating natural recovery in fish and coral populations.** Marine Pollution Bulletin 54(7), 1009-1019.

Read, Charmaine I.; Bellwood, David R.; van Herwerden, Lynne. (2006) **Ancient origins of Indo-Pacific coral reef fish biodiversity: a case study of the leopard wrasses (Labridae: Macropharyngodon).** Molecular Phylogenetics and Evolution 38(3), 808-819.

Richards, Zoe T.; van Oppen, Madeleine J. H.; Wallace, Carden C.; Willis, Bette L.; Miller, David J. (2008) **Some rare indo-pacific coral species are probable hybrids.** PLoS One 3(9)

Richier, Sophie; Cottalorda, Jean-Michel; Guillaume, Mireille M. M.; Fernandez, Cyril; Allemand, Denis; Furla, Paola. (2008) **Depth-dependent response to light of the reef building coral, Pocillopora verrucosa: Implication of oxidative stress.**

Riegl, Bernhard; Bruckner, Andy; Coles, Steve L.; Renaud, Philip; Dodge, Richard E. (2009) **Threats and conservation in an era of global change.** Annals of the New York Academy of Sciences 1162 136-186.

Riitzler, K. and D.L. Santavy. (1983). **"The Black Band disease of Atlantic reef corals. I. Description of the cyanophyte pathogen.".** *P.S.Z.N.I. Mar. Ecol.* 4: 301–319.

Ritchie KB, Smith GW. Type II white-band disease. *Revista De Biologia Tropical.* 1998;46:199–203.

Robbins, William D.; Hisano, Mizue; Connolly, Sean R.; Choat, J. Howard. (2006) **Ongoing Collapse of Coral-Reef Shark Populations.** Current Biology 16(23), 2314-2319.

Roberts, Callum M.; McClean, Colin J.; Veron, John E. N.; Hawkins, Julie P.; Allen, Gerald R.; McAllister, Don E.; Mittermeier, Cristina G.; Schueler, Frederick W.; Spalding, Mark; Wells, Fred; Vynne, Carly; Werner, Timothy B. (2002) **Marine biodiversity hotspots and conservation priorities for tropical reefs.** Science (Washington, DC, United States) 295(5558), 1280-1284.

Rogers, Z. Hillis-Starr, R. Nemeth, M. Taylor, G.P. Schmahl, M.W. Miller, D.A. Gulko, J.E. Maragos, A.M. Friedlander, C.L. Hunter, R.S. Brainard, P. Craig, R.H. Richond, G. Davis, J. Starmer, M. Trianni, P. Houk, C.E. Birkeland, A. Edward, Y. Golbuu, J. Gutierrez, N. Idechong, G. Paulay, A. Tafileichig, and N. Vander Velde (2002) **The State of Coral Reef Ecosystems of the United States and Pacific Freely Associated States:** National Oceanic and Atmospheric Administration/National Ocean Service/National Centers for Coastal Ocean Science, Silver Spring, MD.

Roopin, Modi; Henry, Raymond P.; Chadwick, Nanette E. (2008) **Nutrient transfer in a marine mutualism: patterns of ammonia excretion by anemonefish and uptake by giant sea anemones.** Marine Biology (Heidelberg, Germany) 154(3), 547-556.

Rosenburg E and Loya Y (Eds) (2004) **Coral Health and Disease.** Springer-Verlag, Germany

Rotjan, Randi D.; Lewis, Sara M. (2005) **Selective predation by parrotfishes on the reef coral**

Porites astreoides. Marine Ecology: Progress Series 305 193-201.

S. G. Dove, C. Lovell, M. Fine, J. Deckenback, O. Hoegh-Guldberg, R. Iglesias-Prieto, K. R. N Anthony (2008) **Host pigments: potential facilitators of photosynthesis in coral symbioses** Plant, Cell and Environment 31(11), 1523-1533

S. V. Vollmer, D. I. Kline (2008) **Natural disease resistance in threatened staghorn corals** Marine Science Center, Northeastern University, Nahant, MA, USA.   PLoS One 3(11).

Sekar, Raju; Kaczmarsky, Longin T.; Richardson, Laurie L. (2008) **Microbial community composition of black band disease on the coral host Siderastrea siderea from three regions of the wider Caribbean.** Marine Ecology: Progress Series 362 85-98.

Steven V. Vollmer and David I. Kline (2008) Natural Disease Resistance in Threatened Staghorn Corals. *PLoS ONE* 3(11)

Steven V. Vollmer and Stephen R. Palumbi (2002) Hybridization and the Evolution of Reef Coral Diversity. Science.June 14th. Vol 296. No 5575, pp.2023 - 2025

Sweatman, Hugh (2008) **No-take reserves protect coral reefs from predatory starfish.** Current Biology (2008), 18(14), R598-R599.

Tun, K., L. M. Chou, A. Cabanban, V. S. Tuan, Philreefs, T. Yeemin, Suharsono, K. Sour and D. Lane (2004) **Status of Coral Reefs, Coral Reef Monitoring and Management in Southeast Asia**, . p: 235-276. in C. Wilkinson (ed.). Status of coral reefs of the world: 2004. Volume 1. Australian Institute of Marine Science, Townsville, Queensland, Australia.

V. M. Weis (2008). **Cellular mechanisms of Cnidarian bleaching: stress causes the collapse of symbiosis** Journal of Experimental Biology. 211(19), 3059-3066.

Vollmer, Steven V.; Patumbi, Stephen R. (2002) **Hybridization and the evolution of reef coral diversity.** Science (Washington, DC, United States) 296(5575), 2023-2025.

Voss, Joshua D.; Mills, DeEtta K.; Myers, Jamie L.; Remily, Elizabeth R.; Richardson, Laurie L. (2007) **Black Band Disease Microbial Community Variation on Corals in Three Regions of the Wider Caribbean.** Microbial Ecology 54(4), 730-739

W. S. Fisher, L. S. Fore, A. Hutchins, R. L. Quarles, J. G. Campbell, Jed G, C. LoBue, W. S. Davis (2008) **Evaluation of stony coral indicators for coral reef management.** Marine Pollution Bulletin 56(10), 1737-1745

Webster, Nicole S. (2007) **Sponge disease: a global threat?** Environmental Microbiology 9(6), 1363-1375.

Wood, L. J. (2007). **MPA Global: A database of the world's marine protected areas. Sea Around Us Project, UNEP-WCMC & WWF**. www.mpaglobal.org

Woodman, George H.; Wilson, Simon C.; Li, Vincent Y. F.; Renneberg, Reinhard. (2003) **Acoustic characteristics of fish bombing: potential to develop an automated blast detector.** Marine Pollution Bulletin 46(1), 99-106.

Yap, Helen T. (2004) **Differential survival of coral transplants on various substrates under elevated water temperatures.** Marine Pollution Bulletin 49(4), 306-312.

Z. Dong, H. Huang, L. Huang, Y. Li, R. Zou (2008). **Progress in taxonomy and genetic diversity of Zooxanthellae.** Haiyang Tongbao 27(3), 95-101

Zamzow JP, Losey GS (2002) **Ultraviolet radiation absorbance by coral reef fish mucus: photo protection and visual communication**. Environ Biol Fishes 63:41–47

Zhan, Wenhuan; Sun, Jinlong; Jia, Jianye; Sun, Jie; Yao, Yantao; Liu, Zaifeng; Zhang, Zhiqiang; Zhan, Meizhen; Zhang, Dianguang. (2007) **Environmental changes of volcano activity recorded in coral reefs in the soutwestern Leizhou Peninsula.** Progress in Environmental Science and Technology 1 403-406.

Further reading and websites

World Database on Protected Areas (WDPA): www.wdpa-marine.org
MPA global: www.reefbase.org
Friends of IYOR 2008: www.iyor.org
ARC Centre of Excellence for Coral Reef Studies: www.coralcoe.org.au
Buccoo Reef Trust: www.buccooreef.org
Cape Eleuthera Institute: http://ceibahamas.org
The Coral Reef Alliance: www.coralreefalliance.org
Coralwatch: www.coralwatch.org
Coral Reef Intiatives for the Pacific: www.crisponline.net
The Coral Reef Targeted Research and Capacity Building for Management (CRTR)
Program: http://gefcoral.org/
Florida Reef Resilience Program: http://frrp.org/
Global Coral Reef Monitoring Network (GCRMN): www.gcrmn.org
Great Barrier Reef Marine Park Authority (GBRMPA): www.gbrmpa.gov.au
The International Coral Reef Action Network (ICRAN): www.icran.org
International Coral Reef Research and monitoring Center:
www.coremoc.go.jp/english/top_e.html
IISD Reporting Services: www.iisd.ca
The International Ocean Institute (IOI): www.ioinst.org
National Oceanic and Atmospheric Administration (NOAA): www.noaa.gov
The Partnerships in Environmental Management for the Seas of East Asia (PEMSEA):
www.pemsea.org
Project AWARE: www.projectaware.org
ReefBase - A Global Information System on Coral Reefs: www.reefbase.org
Regional Organization for the Conservation of the Environment of the Red Sea and Gulf of
Aden: www.persga.org
UNEP Caribbean Environment Programme (CEP): www.cep.unep.org
United States All Islands Coral Reef Committee (AIC): http://allislandscorals.org
The Western Indian Ocean Marine Science Association (WIOMSA): www.wiomsa.org
The World Conservation Union (IUCN) Global Marine Programme: www.iucn.org/marine
The World Ocean Observatory: www.thew2o.net/events/index.html

Holly Blue Publishing

www,hollybluepublishing.co.uk

www.ingramcontent.com/pod-product-compliance
Lightning Source LLC
Chambersburg PA
CBHW050730180526
45159CB00003B/1179